Better Homes and Gardens®

STEP-BY-STEP
MASONRY &
CONCRETE

BETTER HOMES AND GARDENS® BOOKS
Editor: Gerald M. Knox
Art Director: Ernest Shelton
Managing Editor: David A. Kirchner

Building and Remodeling Editor: Joan McCloskey
Building Books Editor: Larry Clayton
Building Books Associate Editor: Jim Harrold

Associate Art Director (Managing): Randall Yontz
Associate Art Directors (Creative): Linda Ford,
   Neoma Alt West
Copy and Production Editors: Nancy Nowiszewski,
   Lamont Olson, Mary Helen Schiltz, David A. Walsh
Assistant Art Directors: Faith Berven, Harijs Priekulis
Senior Graphic Designer: Tom Wegner
Graphic Designers: Mike Burns, Alisann Dixon,
   Mike Eagleton, Lynda Haupert, Deb Miner,
   Lyne Neymeyer, Bill Shaw, D. Greg Thompson
Editor in Chief: Neil Kuehnl
Group Editorial Services Director: Duane Gregg
Executive Art Director: William J. Yates

General Manager: Fred Stines
Director of Publishing: Robert B. Nelson
Director of Retail Marketing: Jamie Martin
Director of Direct Marketing: Arthur Heydendael

Step-By-Step Masonry & Concrete
Editors: Larry Clayton, Jim Harrold
Copy and Production Editor: David A. Walsh
Graphic Designers: Tom Wegner, Mike Burns
Contributing Writer: George Brandsberg
Technical Consultant: Lloyd McCabe
Drawings: Jim Stevenson, Carson Ode

Acknowledgments
Our appreciation goes to the following
companies, associations, and individuals
for their help in the preparation of this book:
Brick Institute of America
Masonry Products Division, Cantex Industries
Demco, Inc.
Jack Wilson, Duke Ready-Mix, Inc.
Harvey Sand & Gravel
Kurtz Hardware Co.
Thomas Hornback
Marshalltown Trowel Company
Larry Otto
Bruce Paterson, Executive Director,
   Masonry Institute of Iowa
Portland Cement Association
Rowat Cut Stone Co.
United Brick & Tile Co. of Iowa

CONTENTS

INTRODUCTION

Learning how to do anything (whether it be playing the piano, mastering a foreign language, or in this case working with concrete and masonry) can overwhelm you at first. There's so much to know. By realizing that the whole is simply the sum of the parts, however, the learning can be fruitful and enjoyable.

So as you take up this book in search of a new set of skills, rest assured that we have your needs and your skill level in mind. We've made every attempt to start at the beginning of these interesting subjects and take you step by step through what you need to know to achieve professional-quality results every time.

We begin on page 6 by exposing you to the tools you'll need to do concrete and masonry projects. Then we move on to a visual and verbal presentation of the various materials available to you to use. If you're surprised by all the options shown there, you'll be even more surprised to learn, for example, that there are more than 10,000 different brick styles manufactured, plus many variations of the other materials as well.

And because planning is so vital to the success of any project, we have set aside pages 10-13 for what we call "Planning Pointers." It's here you'll find out how thick to make slabs, how wide to make driveways and sidewalks, and what projects need footings. And we'll answer all those other questions that will almost certainly arise during the planning stage of your projects. A set of specifications, if you will.

Next, we launch into the first of the book's three major sections— "Working with Concrete." You'll learn how to prepare the site for your project, build the forms, get ready for the pour, estimate your concrete needs, and mix or order concrete. We follow this with information about how to place, finish, and cure concrete as well as how to perform several

concrete repairs. We're confident that by the time you've digested the information in this section you'll be able to handle any residential concrete project—and handle it with skill.

In the book's second section—"Working with Block, Brick, and Stone" —we share with you the information and the techniques you need to become a proficient amateur mason. From how to estimate your materials needs, to laying masonry units without mortar, to laying

up walls of various types and materials, to making some common masonry repairs, it's all there—in simple, easy-to-follow fashion.

Then on page 74 begins a 19-page section— "Special-Effect Projects" —in which you find 10 exciting creations you may want to build or use as inspiration.

And to round out the book and to help you find things fast, we conclude with a glossary and a thorough index.

Tools for Masonry and Concrete

Nothing can substitute for having the right tool for a job. And concrete and masonry projects are no exceptions. Here's a brief rundown on the tools of the trade you need to get your job done.

As you read, keep in mind you can either buy or rent all of these items—whichever better suits your immediate and future needs. If you buy them, purchase only good-quality tools; bargain-bin items simply won't hold up.

You may already have some of the tools necessary for the tasks this book deals with—a *claw hammer*, *handsaw* or *portable circular saw*, *carpenter's framing square*, and a flexible *tape measure*. These, along with a *4-foot-long level*, *line level* and *mason's line*, and a *chalk box*, will come in handy when building wooden forms for containing concrete, among other jobs.

To aid in mixing and transporting masonry materials and concrete to the job site, lay your hands on a sturdy *contractor-quality wheelbarrow*—the kind with a 3-cubic-foot tray (or larger) and a pneumatic rubber tire. Also handy to have but not essential are a *mortar box* and *mortar hoe* for mixing ingredients.

To prepare sites and move concrete and mortar ingredients, you must have either a round-bladed shovel or a couple of specialists—a *square-bladed shovel* for moving sand and wet concrete, and a *spade* for touching up a slab excavation or footing trench.

For placing concrete, you'll need a *tamper*, a *screed* (not shown), and a *darby* or *bull float*. Most people simply employ a straight length of 2×4 (as a screed) to level just-poured concrete. Use the darby, bull float, or a small hand float to smooth up screeded concrete and to

push larger pieces of stone below the surface. You can make your own tamper, darby, and floats from scrap lumber, if desired.

To finish concrete, get a magnesium or steel trowel to further smooth and compact the concrete, an inexpensive *edger* to round off and strengthen the edges of concrete slabs, and a *jointer* to put grooves in slabs to control cracking. If you plan to cut control joints in the concrete after placing and finishing it, go with a circular saw that has a *masonry blade* for the jointing work.

And if you plan to finish the concrete with a nonskid surface, you'll want to have handy a *stiff-bristled push-type broom* (not shown). (Brooming takes the place of steel troweling when a rougher surface is desired.)

To cure the concrete, all you need is a garden hose with an adjustable nozzle or a lawn sprinkler that oscillates.

You can make cutting bricks, blocks, or stones much easier by having handy a *brick hammer*, a *brick set* or *masonry chisel*, and a 2-pound *sledge hammer* (sometimes called a mash hammer).

And when building concrete or masonry walls, you'll need *line blocks*, a *modular spacing rule*, and a *plumb bob*.

For placing mortar, use a well-balanced, pointed *brick trowel*. Use a *pointing trowel* or a *caulking trowel* to force mortar into joints being repaired. A *joint strike* or a *joint raker* (not shown) helps you "strike" (finish) mortar joints between bricks, blocks, or stones.

A note about cleanup: After you finish with any concrete or masonry tool, wash the implement thoroughly, and then dry it. Follow this with a light coat of oil on all metal parts. Why? Dried concrete and mortar are almost impossible to remove.

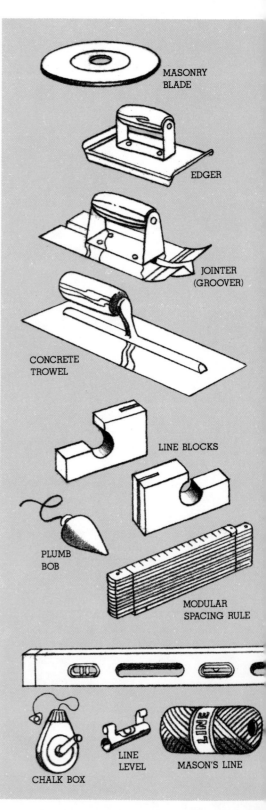

MASONRY BLADE

EDGER

JOINTER (GROOVER)

CONCRETE TROWEL

LINE BLOCKS

PLUMB BOB

MODULAR SPACING RULE

CHALK BOX

LINE LEVEL

MASON'S LINE

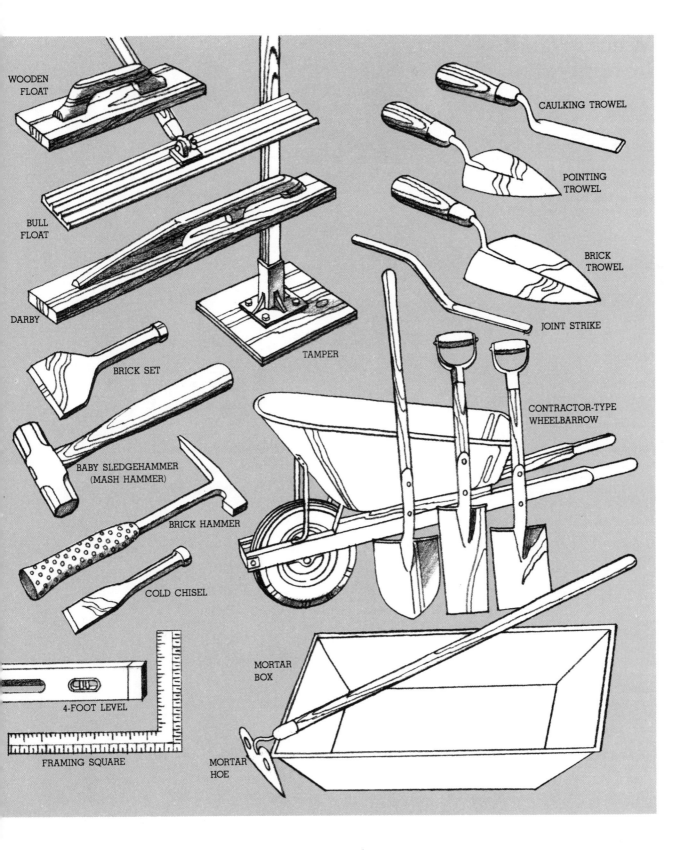

WOODEN FLOAT

CAULKING TROWEL

POINTING TROWEL

BULL FLOAT

BRICK TROWEL

DARBY

TAMPER

JOINT STRIKE

BRICK SET

CONTRACTOR-TYPE WHEELBARROW

BABY SLEDGEHAMMER (MASH HAMMER)

BRICK HAMMER

COLD CHISEL

4-FOOT LEVEL

MORTAR BOX

FRAMING SQUARE

MORTAR HOE

8

What's What in Materials

Masonry materials come in an amazing variety of sizes, shapes, colors, and other qualities. Luckily, the trade has adopted a handful of standard items to make it easier to choose what you need for a particular job.

Concrete, one of the most widely used building materials, is nothing more than a correctly proportioned mixture of coarse aggregate (gravel or crushed rock), sand, cement, and water. When combined, these ingredients quickly transform into a stone-like mass.

Mortar (a paste made of water, cement, lime, and sand) binds bricks, concrete blocks, and stones much the same way cement, sand, aggregate, and water work in concrete. In addition, mortar serves as an attractive spacer between materials and fills in around (and helps hide) their imperfections.

Bricks, most of which are manufactured by firing molded clay or shale, vary widely in color, texture, and dimensions. All fall, however, into four main categories: *common* or *building, patio, fire,* and *facing.*

You also should be aware of two other designations that exist in the world of bricks. *Modular bricks* are sized to conform to modules of 4 inches, which makes estimating material needs quite predictable. *Non-modular bricks* do not conform to the 4-inch module.

For outdoor projects that must withstand moisture and freeze-thaw cycles, ask for SW (severe weathering) grade bricks. For indoor uses, such as facing a fireplace or a planter, you can use MW (moderate weathering) or NW (no weathering).

When a dealer tells you the dimensions of a brick, he is talking about its *nominal size,* which is the actual size of the brick plus a normal mortar joint of ⅜ to ½ inch on the bottom and at one end.

Concrete blocks are cast from a stiff concrete mix. A typical *stretcher* block has a nominal size of 8×8×16 inches and weighs 40 to 50 pounds. Hollow cores in the block help conserve material, make the blocks easier to grip and place, add insulation value, and provide channels for utilities.

Concrete blocks come in two different grades—*N,* designed for outdoor freezing and thawing conditions, and *S,* for use above grade and where it will not be exposed to the weather directly. These blocks come in hundreds of sizes and shapes (see drawing at right for a small sampling).

You also can purchase concrete blocks that have decorative patterns molded into them, as well as a wide selection of screen blocks, which come in many sizes.

Glass blocks, an anomaly here, come in square 6-, 8-, and 12-inch nominal sizes, all 3⅞ inches thick. They're especially useful for making basement windows and for providing outside light in entry areas and bathrooms.

Stone—you encounter three types: *rubble* (round rocks), *flagstone* (flat irregular pieces), and *ashlar* or dimensioned stone, which is cut into slices of uniform thickness for laying in coursed or non-coursed patterns.

Lightweight veneers of brick, terra cotta, or stone (either natural or artificial) are useful for putting an attractive outside layer on otherwise plain materials such as poured concrete, concrete blocks, or wood-framed, sheathed walls.

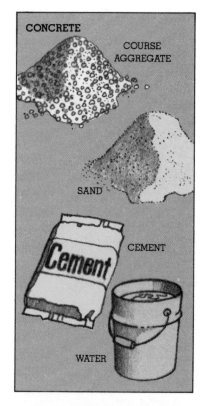

CONCRETE
COURSE AGGREGATE
SAND
CEMENT
WATER

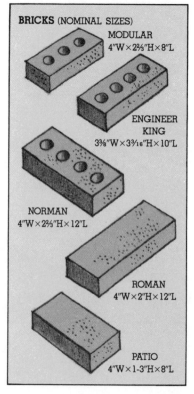

BRICKS (NOMINAL SIZES)
MODULAR 4"W×2⅔"H×8"L
ENGINEER KING 3⅝"W×3¹⁄₁₆"H×10"L
NORMAN 4"W×2⅔"H×12"L
ROMAN 4"W×2"H×12"L
PATIO 4"W×1-3"H×8"L

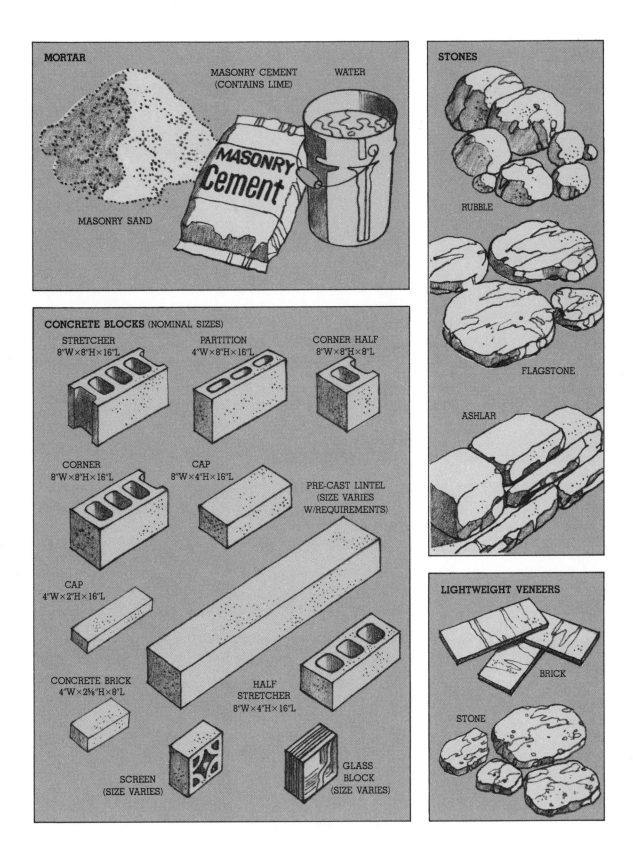

MORTAR

MASONRY CEMENT
(CONTAINS LIME)

WATER

MASONRY
Cement

MASONRY SAND

STONES

RUBBLE

FLAGSTONE

ASHLAR

CONCRETE BLOCKS (NOMINAL SIZES)

STRETCHER
8″W×8″H×16″L

PARTITION
4″W×8″H×16″L

CORNER HALF
8″W×8″H×8″L

CORNER
8″W×8″H×16″L

CAP
8″W×4″H×16″L

PRE-CAST LINTEL
(SIZE VARIES
W/REQUIREMENTS)

CAP
4″W×2″H×16″L

CONCRETE BRICK
4″W×2⅝″H×8″L

HALF
STRETCHER
8″W×4″H×16″L

SCREEN
(SIZE VARIES)

GLASS
BLOCK
(SIZE VARIES)

LIGHTWEIGHT VENEERS

BRICK

STONE

Planning Pointers

Throughout this book, you'll run across statements such as "Take the time to do (this or that)" and "Take care that you do (this or that)." That's because with masonry and concrete, the degree of care you use to perform various operations (and pay attention to detail) translates into a good-looking result or one that's unsatisfactory.

As important as craftsmanship is to a masonry or concrete project, however, it pales in comparison to the planning considerations involved. It's almost impossible to get satisfactory results without a thoughtful plan. That's why we spend four pages of the book on starting out right.

Following are some generalities you should keep in mind when building with concrete or masonry, and on pages 12–13 we show you the planning specifics—answers to those questions we know you'll have as you progress through the planning stage of any of several projects.

Make It Functional

Beauty alone can't make a project successful. No, it must serve a useful function, too. So you must first ask yourself what you want the project to do, then go about trying to figure out the best way to accomplish your aims.

Let's just take a concrete slab patio, for example (the process is the same no matter what the project). With patios, size is definitely a factor; you should allow 20 square feet times the number of persons you expect to have on it at one time.

Also consider access. For this project to work as a logical extension of the house, you must have easy access to it from an outside house entrance.

Climate and weather, too, have a bearing on where your patio goes. You may want to avoid southern or western exposures where you would bake under the summer sun. Prevailing winds, rain, and winter conditions all figure in, too.

Follow the Rules

Most communities enforce zoning regulations and building codes. So, whenever you want to build an outbuilding, attach an addition to your home, or make any other changes to your property, first obtain a building permit (this is especially true in incorporated areas).

Building codes set the minimum standards for constructing a project. For example, a code may dictate the dimensions and type of concrete you must use for footings or acceptable methods of construction. They also usually spell out the "setback," which is how close to your property line you may build.

Inspect the deed to your property, too. It may include restrictive covenants or easements or other uses that will affect what you can build and where you can put it.

Find out the location of all underground utilities on your property and avoid placing structures over underground septic systems or gas, water, electric, or telephone lines.

Show Some Style

Short of hiring an architect, the best way to get ideas and a sense of style is to drive around town and see what other people have done. If you see something you like, stop in and chat with the owners. And keep your eyes open for possibilities as you flip through the pages of your favorite shelter magazine or book. Also look over the projects in this book. And remember, the simplest designs are often the best looking.

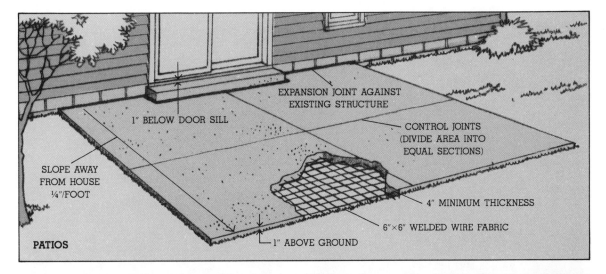

EXPANSION JOINT AGAINST
EXISTING STRUCTURE

1" BELOW DOOR SILL

CONTROL JOINTS
(DIVIDE AREA INTO
EQUAL SECTIONS)

SLOPE AWAY
FROM HOUSE
¼"/FOOT

4" MINIMUM THICKNESS

6"×6" WELDED WIRE FABRIC

PATIOS

1" ABOVE GROUND

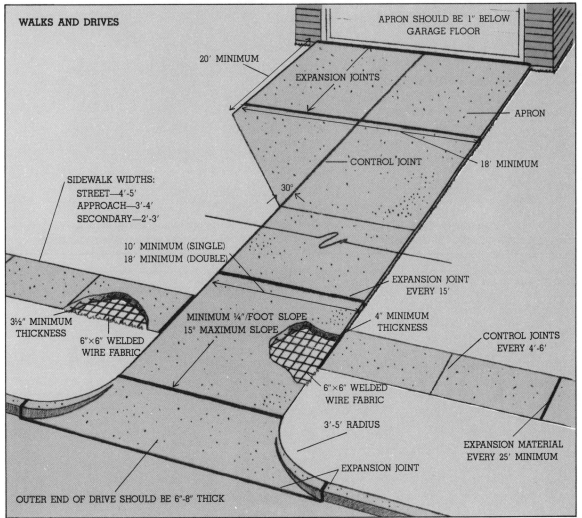

WALKS AND DRIVES

APRON SHOULD BE 1" BELOW
GARAGE FLOOR

20' MINIMUM

EXPANSION JOINTS

APRON

CONTROL JOINT

18' MINIMUM

SIDEWALK WIDTHS:
STREET—4'-5'
APPROACH—3'-4'
SECONDARY—2'-3'

30°

10' MINIMUM (SINGLE)
18' MINIMUM (DOUBLE)

EXPANSION JOINT
EVERY 15'

3½" MINIMUM
THICKNESS

6"×6" WELDED
WIRE FABRIC

MINIMUM ¼"/FOOT SLOPE
15° MAXIMUM SLOPE

4" MINIMUM
THICKNESS

CONTROL JOINTS
EVERY 4'-6'

6"×6" WELDED
WIRE FABRIC

3'-5' RADIUS

EXPANSION MATERIAL
EVERY 25' MINIMUM

EXPANSION JOINT

OUTER END OF DRIVE SHOULD BE 6"-8" THICK

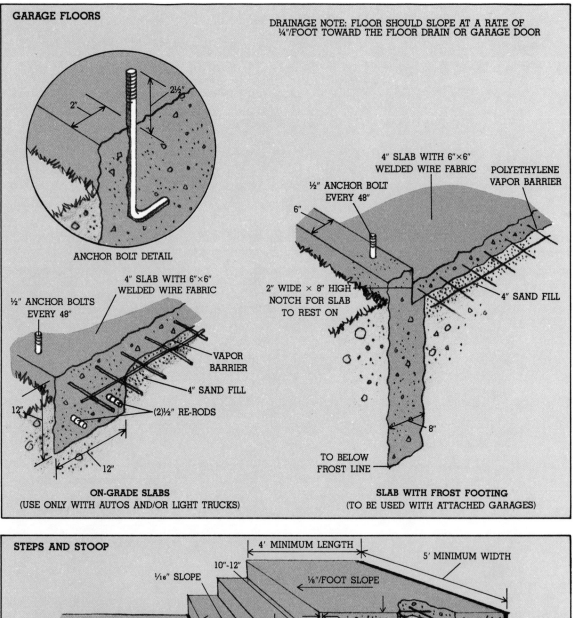

GARAGE FLOORS

DRAINAGE NOTE: FLOOR SHOULD SLOPE AT A RATE OF
¼"/FOOT TOWARD THE FLOOR DRAIN OR GARAGE DOOR

2"

2½"

ANCHOR BOLT DETAIL

4" SLAB WITH 6"×6"
WELDED WIRE FABRIC

POLYETHYLENE
VAPOR BARRIER

½" ANCHOR BOLT
EVERY 48"

6"

4" SAND FILL

2" WIDE × 8" HIGH
NOTCH FOR SLAB
TO REST ON

8"

TO BELOW
FROST LINE

½" ANCHOR BOLTS
EVERY 48"

4" SLAB WITH 6"×6"
WELDED WIRE FABRIC

VAPOR
BARRIER

4" SAND FILL

12"

(2)½" RE-RODS

12"

ON-GRADE SLABS
(USE ONLY WITH AUTOS AND/OR LIGHT TRUCKS)

SLAB WITH FROST FOOTING
(TO BE USED WITH ATTACHED GARAGES)

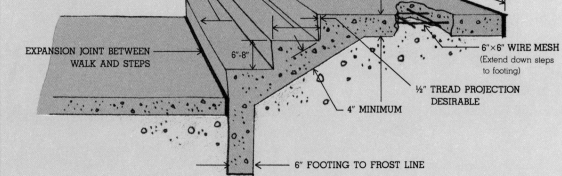

STEPS AND STOOP

4' MINIMUM LENGTH

5' MINIMUM WIDTH

10"-12"

1/16" SLOPE

⅛"/FOOT SLOPE

EXPANSION JOINT BETWEEN
WALK AND STEPS

6"-8"

6"×6" WIRE MESH
(Extend down steps
to footing)

½" TREAD PROJECTION
DESIRABLE

4" MINIMUM

6" FOOTING TO FROST LINE

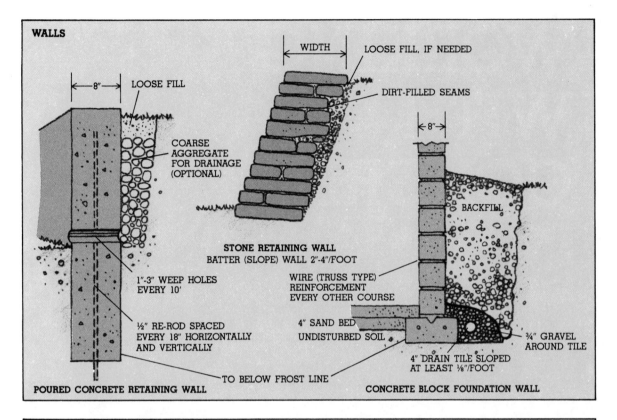

WALLS

8″ — LOOSE FILL

WIDTH — LOOSE FILL, IF NEEDED

DIRT-FILLED SEAMS

COARSE AGGREGATE FOR DRAINAGE (OPTIONAL)

8″

BACKFILL

STONE RETAINING WALL
BATTER (SLOPE) WALL 2″-4″/FOOT

1″-3″ WEEP HOLES EVERY 10′

WIRE (TRUSS TYPE) REINFORCEMENT EVERY OTHER COURSE

½″ RE-ROD SPACED EVERY 18″ HORIZONTALLY AND VERTICALLY

4″ SAND BED
UNDISTURBED SOIL

¾″ GRAVEL AROUND TILE

4″ DRAIN TILE SLOPED AT LEAST ⅛″/FOOT

TO BELOW FROST LINE

POURED CONCRETE RETAINING WALL

CONCRETE BLOCK FOUNDATION WALL

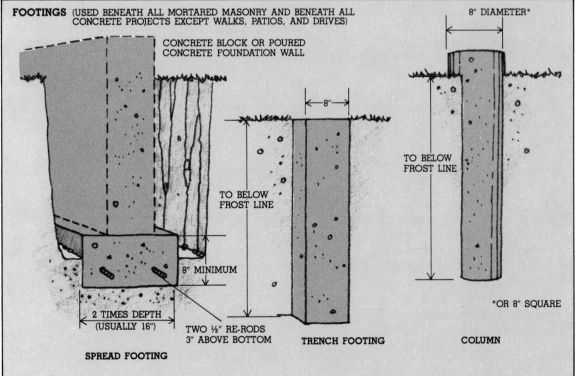

FOOTINGS (USED BENEATH ALL MORTARED MASONRY AND BENEATH ALL CONCRETE PROJECTS EXCEPT WALKS, PATIOS, AND DRIVES)

8″ DIAMETER*

CONCRETE BLOCK OR POURED CONCRETE FOUNDATION WALL

8″

TO BELOW FROST LINE

TO BELOW FROST LINE

8″ MINIMUM

2 TIMES DEPTH (USUALLY 16″)

TWO ½″ RE-RODS 3″ ABOVE BOTTOM

*OR 8″ SQUARE

SPREAD FOOTING

TRENCH FOOTING

COLUMN

WORKING WITH CONCRETE

Concrete is a marvelous material. Its strength, durability, and ease of handling make it one of the most popular materials for construction jobs ranging from paving walkways to erecting skyscrapers.

The most exciting thing about concrete, however, is that YOU can use it too. In fact, there's a whole concrete industry eager to help you get started on even small projects.

If the prospect of handling, mixing, and placing large piles of sand and gravel and heavy bags of cement overwhelms you, relax. Don't let the thought of pouring possibly tons of mush that turns rock hard in a few hours scare you. Just think of concrete as measured amounts of sand and gravel pasted together with wet cement.

In that light, it is very ordinary. A small batch, as you'd use for setting a post, is easy to mix, to place, and to finish.

As you increase the size of the project, you need more of everything —planning, materials, and helpers.

Success depends on doing the right thing when you should. The old saying "Plan your work and work your plan" has special meaning when you undertake a concrete project. For planning basics, see pages 10–13. Once you've settled on a plan, you're ready to prepare the site.

Bear in mind that working with concrete is like climbing stairs: Take one step at a time in the proper order and you're sure to come out on top.

Preparing the Site for Slabs

Though Samson-like in many respects, concrete does have a weakness: its tensile (lateral) strength. A slab without a firm, uniform base will almost certainly crack and heave, leaving you with a none-too-easy repair. This can happen with driveways where the sand base underneath washes away, and cracks develop under the weight of vehicles.

If at all possible, place the slab on undisturbed soil. However, if you'll be pouring on top of recently filled soil, as would be the case with backfill around a new foundation, pack the soil by watering it for several days and allowing it time to settle. Compact small areas of loose dirt by moistening and tamping them.

To provide for drainage on low or boggy sites, line the base with two or more inches of gravel or sand. And to ensure that rain will run off of the slab, allow a slope of ¼ inch per foot.

1 Stake the outline of the proposed slab and square the corners, using the 3–4–5 method shown. Any multiple of this ratio, such as 6–8–10 or 9–12–15, will give you the same results. To make sure the perimeter is square, measure the diagonals between opposite corners. If the distances are equal, the corners are square. *(continued)*

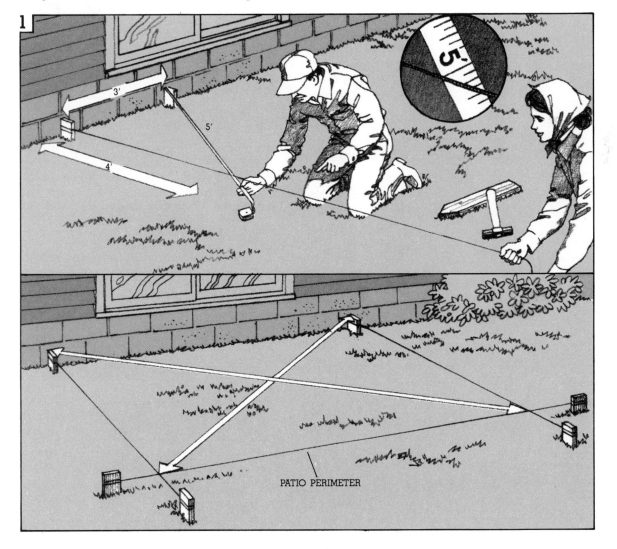

PATIO PERIMETER

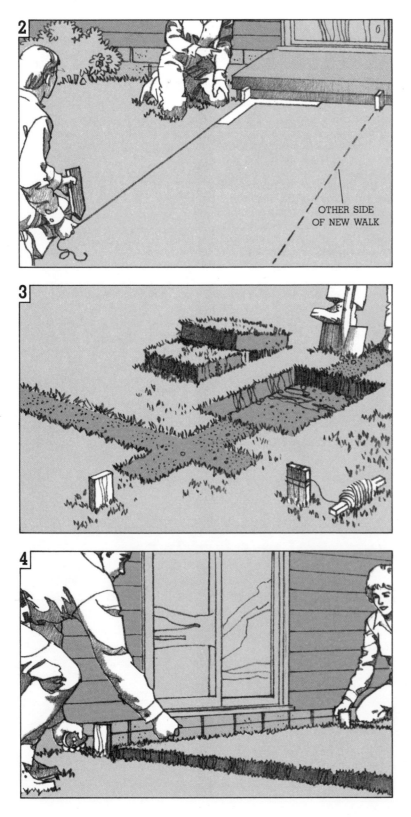

OTHER SIDE
OF NEW WALK

Preparing the Site for Slabs *(continued)*

2 To locate one side of a long, narrow slab such as a walk, use a carpenter's framing square. Locate the opposite side by measuring off the desired width.

3 Mark the location of the slab edges by sprinkling sand over the mason's line. Then remove the line and dig a shallow trench that reaches horizontally about 3 inches beyond the edge of the slab.

4 After trenching all the way around the perimeter, snap a chalkline on the existing structure to establish the surface of the slab where it abuts the foundation. Place the mark high enough to allow plenty of slope away from the building. (Note: The surface of an outdoor slab should be about 1 inch above final ground level.)

5 Now adjust the mason's line for proper slope (¼ inch/foot) and drive 1×4 stakes so the tops match the level of the line. Be sure to place the stakes outside of the line by a distance equal to the thickness of the forming lumber you plan to use.

6 Remove the sod and enough top soil to reach the desired depth. If you plan to put a layer of sand or gravel beneath the slab, excavate at least 5 inches deep. As you dig, check the bottom from time to time by laying a 2×4 or 2×6 on edge so its top barely touches the mason's line. If you find you've removed too much soil in some spots, fill them later with sand or gravel rather than with loose soil.

A flat spade works best for shaving away the final inch or so of soil from the bottom and sides of the excavation.

When excavating where you have well-established turf, you may want to take up the sod for re-laying elsewhere. Undercut it horizontally about 2 inches beneath the surface and cut it into easy-to-handle 8×16-inch sections. Save enough to re-sod around the edges of your new slab, too.

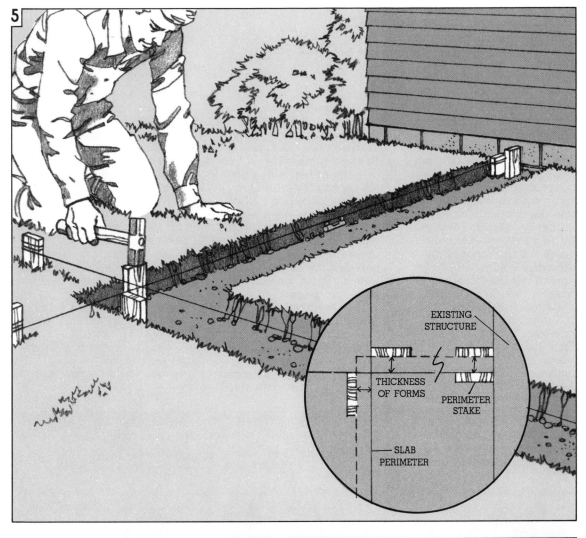

EXISTING
STRUCTURE

THICKNESS
OF FORMS

PERIMETER
STAKE

SLAB
PERIMETER

18

Preparing the Site for Projects Requiring Footings

Because their weight is broadly distributed over many square feet, concrete slabs tend not to need footings. This characteristic allows slabs to "ride out" any frost heave that may occur.

Not so with heavy structures such as concrete walls and any project bonded together by mortar. These require a footing—normally twice the thickness of the structure it supports—to spread out the burden so it doesn't sink. The footing must reach below the frost line to avoid possible damage from frost heaving. (Note: Frost lines vary from place to place; ask local building officials or concrete distributors for the depth in your area.

To be permanently solid, a footing must rest on undisturbed soil. And in the case of foundation walls, proper drainage is a must as well. To ensure this, place a line of clay tile or perforated plastic pipe along the footing to carry seepage into outlets that open at a lower elevation. Then surround the tile with a layer of ¾-inch gravel or crushed rock. Doing this prevents the drainage system from clogging with debris or mud. (See page 13.)

The sketches here show what's involved in preparing a site for a large project requiring footings, but the same basic approach applies to smaller jobs such as garden walls or concrete steps.

1 Lay out the perimeter of the project by driving stakes at the corners. (For details on setting perimeter stakes, see page 15.) About 4 feet outside of the stakes, drive 1×4 stakes and nail on *batter boards* as shown. Because the trench must be at least 3 feet wide to allow room to work, your batter boards should be 4 to 5 feet long.

Now transfer the building lines (A) to the batter boards. This is a two-person operation. Have someone dangle a plumb bob over the outer edge of each perimeter stake while you stretch a length of mason's line as shown. Mark the point at which the lines intersect on the batter boards.

With the building lines transferred, measure over the distances shown in the detail to record the

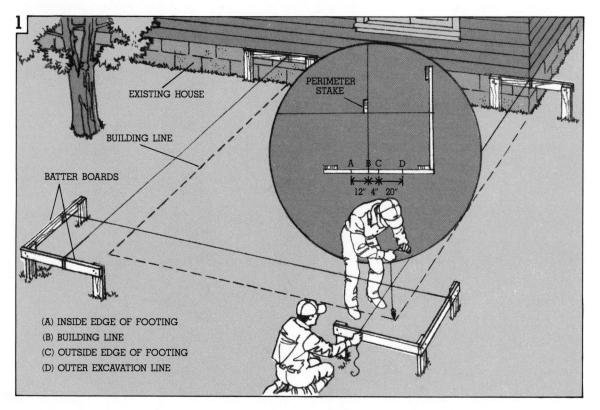

EXISTING HOUSE

BUILDING LINE

BATTER BOARDS

PERIMETER STAKE

A B C D
12" 4" 20"

(A) INSIDE EDGE OF FOOTING
(B) BUILDING LINE
(C) OUTSIDE EDGE OF FOOTING
(D) OUTER EXCAVATION LINE

position of (B) the *inside edge of the footing*, (C) the *outside edge of the footing*, and (D) the outer *excavation line*. Saw kerfs in the batter boards to establish each of the lines mentioned.

2 After digging the trench for the footing, use a plumb bob to locate the outside corners of the footing. Drive stakes (partway in) that line up with the outside edges of your forming lumber (see the detail). Follow the same procedure to position the form stakes for the inside edge of the footing form.

3 Drive form stakes adjacent to the existing footings (if any) so the tops of the stakes and footing are at the same level. Then drive the outside corner stakes so they are level with those next to the building. Check for level with a line level or a carpenter's level supported by a length of straight 2×4 or 2×6 on edge. (With projects that won't be connected to an existing structure, pound the stakes into the ground until the tops of the stakes match the determined height of the footing.)

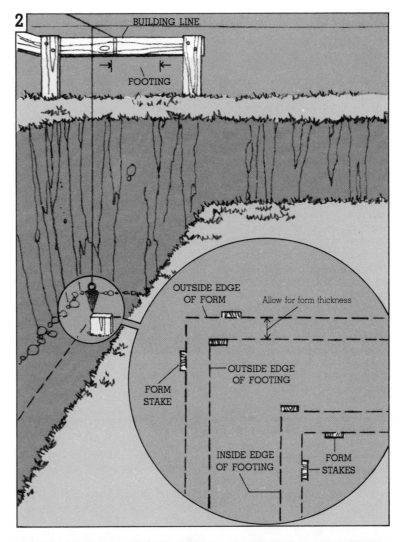

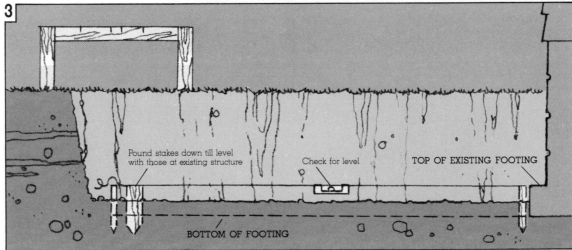

Building Forms

"Cast in concrete" is a way of saying that something is permanent, practically impossible to change. Remember, too, that as wet concrete flows into a form, it fills the niches and faithfully reproduces every detail of the mold you provide.

If you build strong forms that are straight where they should be, or curved if that's part of the plan, the final product will look professional. But if you put up forms that bulge, tilt, or have loose-fitting joints, the finished product will be around for years to come showing off every embarrassing flaw. So when building forms, take a little extra time to be fussy. Inspect your forming lumber for knotholes, cracks, and other defects that you don't want to appear later in the final product.

As a rule, forms for any concrete project must be sturdy, straight, and plumb. If you're in doubt about whether the forms are rigid enough, drive in an extra stake or two and install added braces to keep them in place.

For some jobs, such as the forming for footings, you may be able to excavate carefully and use the soil walls for forms instead of making them with lumber. The key here, of course, is having soil that is firm enough so the excavation holds its shape when you place concrete in it.

Slab Forms

1 Because slabs generally run about 4 inches thick, 2×4s that are smooth and straight make an ideal forming material. When anchoring the forms, do so by driving two double-headed nails through the stakes and into the 2×4s. Make sure to align the top of the forms with that of the stakes. Buttress each form with foot-long 1×4 stakes every 3 to 4 feet. (You also can cut stakes from 1×2 or 2×2 material or buy them pre-cut from your building supplier.) If you opt for 1×4 forming members rather than 2×4s, you'll need to drive stakes every 2 to 3 feet along them to provide adequate support.

2 Do your plans call for a few curves and rounded corners? No problem! Simply substitute 3½-inch-wide strips of ¼-inch-thick hardboard or plywood of suitable length instead of using lumber. If you go with plywood, cut the strips perpendicular to the wood grain of the surface plies so they will be easier to bend. For a snug fit, tack one end in place, in this case driving two 4-penny nails through the thin form member and into the stake. Now spring the materi-

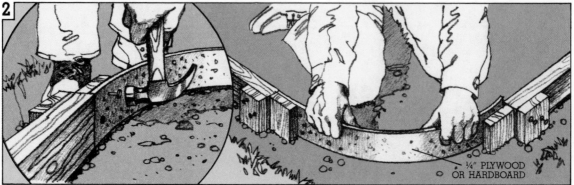

¼" PLYWOOD
OR HARDBOARD

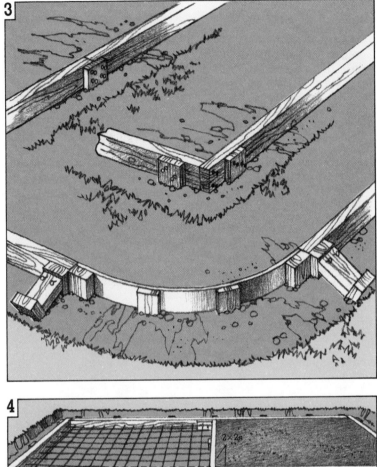

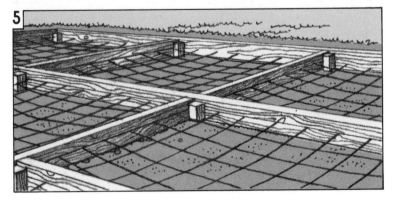

al into the shape you want, mark the point where you'll cut it, and saw off the excess material before completing the installation.

3 When bracing your forms, pay particular attention to wherever two forms meet. For those that butt end to end, drive stakes that lap the joint. For corner situations, station a stake near the end of each form. And to strengthen curved forms, place 1×4 stakes every 1 to 2 feet along the outside radius of the curve.

4 Dividing a large project such as a wide patio or driveway into smaller slabs permits you to pour a manageable amount of concrete at a time. Before beginning, decide whether you want the divider forms to be temporary or permanent. You can use any straight length of lumber for a temporary form; if you plan to leave them in, however, choose redwood, cypress, cedar, or pressure-treated wood for long-lasting good looks. Note here that the 2×2 divider forms sandwich the reinforcing mesh, keeping the mesh at the appropriate level for maximum effectiveness. Drive stakes every 2 to 3 feet to ensure adequate support.

5 If you've elected to use permanent divider forms, apply a coat of wood sealer to further enhance the natural rot resistance of the wood. Then put masking tape on the top surfaces to keep wet cement from staining the wood and to avoid scratches. Drive inside form stakes 1 inch below the top of the form so that when concrete fills the form only the decorative form rails can be seen.

Step Forms

Stairs serve an admittedly utilitarian but vital function. They divide vertical differences into steps. Each step consists of a horizontal *run* and a vertical *rise*. For adults, a rise of about 7 inches is both comfortable and safe.

Here's a useful stair-planning rule: The sum of the run and rise should approximate 18 inches. So, if you want a series of steps with a gentle rise of, say, 5 inches, you should provide runs (or treads) about 13 inches deep. Be sure to use uniform rises and runs in each set of steps. Unexpected changes can cause accidents.

To decide how many steps you need, measure the total rise between levels and divide it by 7 inches (or any other rise you wish to use).

Round off the result to the nearest whole number to get the number of steps. Divide total rise by the number of steps to get the exact rise per step. (To determine the run of each step, repeat the simple math outlined above.)

CAUTION: Local building codes and restrictions may dictate such dimensions as size of risers and treads and the relationship between rises and runs, width, height of sets of steps, landing size, and footing requirements. So check with local building authorities to ensure your plans conform before building.

In general, steps leading to a home should be 4 feet wide or at least as wide as the door and walk they adjoin. For flights rising more than 5 feet, a landing with at least a 3-foot run is desirable. A landing beneath an exterior doorway, however, should ex-

tend 1 foot on each side of the opening and have a run that is 5 feet.

Determining the critical dimensions of rise, run, stair width, and landing size is the hardest part of getting ready to build step forms. But doing this chore with care will make the rest of the job easy.

For flights of more than two steps, you should provide frost-free footings (see page 12 for all the details).

1 Onto an appropriate-size piece of ¾-inch-thick exterior-grade plywood you'll use for one of the form side-walls, draw lines showing total rise and total run (be sure to allow for the 1½-inch bottom riser form as you make the total run marks.) Next, mark the end of the landing and draw lines establishing the location of the finished treads and risers. Note, too, that you'll want the landing to slope away from the house at a rate of ¼ inch per foot of run.

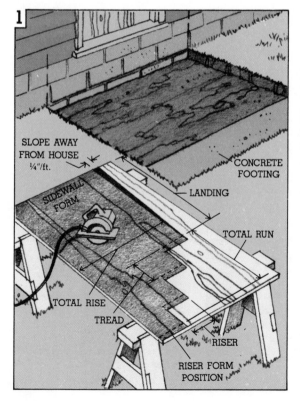

SLOPE AWAY FROM HOUSE ¼"/ft.

SIDEWALL FORM

CONCRETE FOOTING

LANDING

TOTAL RUN

TOTAL RISE

TREAD

RISER

RISER FORM POSITION

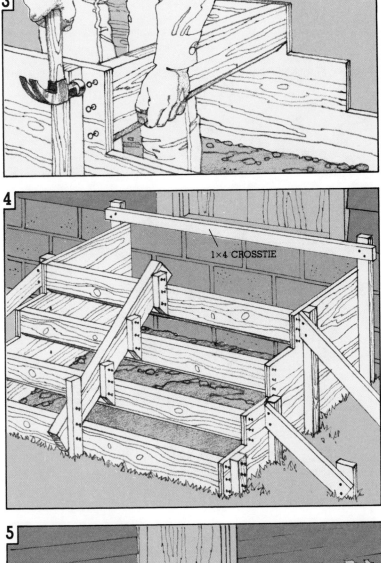

1×4 CROSSTIE

2 After cutting both sidewalls, position them at the entry, using a carpenter's square to make sure they are perpendicular to the building wall. Check the forms for proper slope and plumb before nailing them to support stakes. To keep from loosening the stakes, back them with a heavy maul while driving the nails.

3 Cut riser forms of the correct height and length. Bevel the bottom of the outside of each riser (except the bottom one), leaving about ⅛-inch unbeveled thickness for strength. Cutting this bevel makes it easier to work the concrete trowels and properly finish the tread after you place the concrete. Install the top riser first and the bottom one last to avoid putting unnecessary loads on them.

4 For stairs wider than 4 feet, reinforce the riser forms by driving a 2×4 stake near the center of the bottom form and nailing to it a piece of 2×6 or 2×8 to anchor cleats that prevent riser forms from bulging. To shore up the sidewall forms, drive 2×4 stakes into the ground, then nail 1×4 braces to the 2×4s and to the stakes anchoring the sidewall forms. And finally, to prevent the weight of the concrete from forcing the sidewalls apart, nail on a 1×4 crosstie.

5 To erect a form for concrete steps that parallel the building, first strike a level line on the building wall to establish the desired landing height. Measure down from this line to the level excavation below for the rough height of the plywood back and side members. Don't forget to factor in ¼-inch slope per foot when drawing the sidewall patterns. Also review step 1 for how to figure rise and run. Now cut and install the members, squaring the corners as you go. Next, cut and install a nailer strip diagonally as shown. Attach beveled vertical cleats to the riser forms, then secure the riser forms to the sidewall form and the diagonal nailer, remembering to square and slope them. Firm up the risers with horizontal cleats.

Wall Footing Forms

To be adequate, a wall footing must meet three rules of thumb you need to follow when forming it:

First, in areas where winters are cold enough to freeze the ground, it must rest in an excavation that reaches below the local frost line. Ask your concrete supplier how deep this line is in your locality. Also check with local building authorities to see whether you need a building permit to erect a structure requiring footings and whether local codes spell out any dimension requirements.

Second, the footing must be wide enough to carry the weight of the wall. In general, this means a footing should be twice the width of the wall it supports.

Third, it must be thick enough so it will not fail. As a rule, the thickness of a footing should equal the width of the wall, but be no less than 8 inches thick.

One other pointer: To protect against overhead structural damage that could result when a footing cracks and shifts because of unstable soil conditions, use two or three ½-inch reinforcing rods the full length of the footing. See page 29 for more on reinforcement specifics, and page 13 for information on wall footing drainage.

1 Except for "step footings," which we discuss in caption 4, the only forming materials you'll need are 2×4s for the footing rails and 1×4 stakes. First position the stakes (see page 16 for how to do this). Then secure the rails to them, making sure the top of the rails aligns with the top of the stakes.

If the soil is firm enough to hold its shape when filled with wet concrete, you can dispense with the lumber

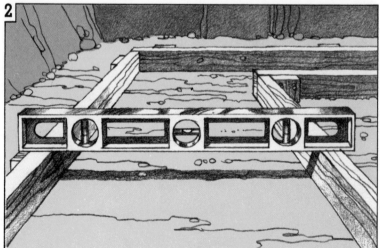

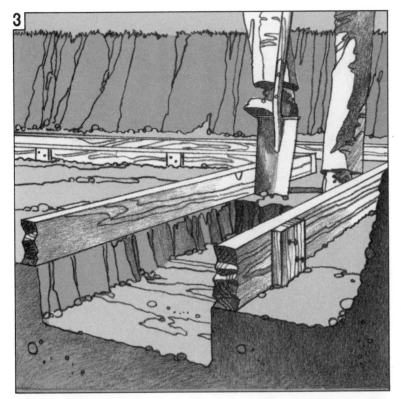

altogether and fabricate a *trench footing.* Simply excavate to the proper dimensions, taking care to maintain straight walls and correct form shape. Check the depth of the trench form, then drive a row of stakes down the center, about 4 feet apart to serve as screeding guides. Check the stakes for level by stretching a line between the end stakes and by hanging a line level in the center.

2 Use a carpenter's level to make sure parallel forms are the same height every few feet along the length of the form. Also check to see that the forms are level lengthwise. Driving stakes every 4 feet should anchor the forms securely. Make sure the stakes you use reach at least 6 inches below the bottom of the trench so form boards will not shift when excavation begins.

3 Once you're satisfied that the forms are level and secure, excavate an additional 4 or so inches of earth. Carefully shape the soil to the proper form dimensions with a spade.

4 To save concrete, you can "step" footings that follow a slope. Shown here are two ways to go. With the first you need 2×8s and stakes, whereas the second uses an earth form and a wooden riser wedged securely between the sidewalls. Note: Stepped forms should rise no more than 2 feet per step and the upper and lower forms should overlap no less than 2 feet.

Poured Concrete Wall Forms

Forms for poured walls must be much stronger than those for most other concrete projects. Why? Because they must be able to counteract two substantial forces: the weight of the material itself and the hydrostatic pressure related to the height of the form and the rate at which you pour the concrete.

Of course, if you mix your own concrete, you can pour it more slowly and use somewhat less massive forms. But for ready-mix concrete that's delivered directly from the chute of a truck, be sure to provide plenty of form support through bracing and by using the right material for the job. Use ¾-inch plywood backed by 2×4 studs every 24 inches for a sturdy concrete wall mold.

1 Sidewall forms are really nothing more than plywood-faced 2×4 stud walls. Working on a flat surface, nail the 2×4 frame together with two 16-penny common nails at each joint, then fasten plywood sheathing with a few 8-penny common nails around the perimeter. Coat the inside form surface with new or used motor oil before setting the form in place. Nail end caps securely.

2 Horizontal *walers* made of doubled 2×4s add strength to the form. Secure them to each sidewall at intervals of no more than 3 feet. Drill ⅛- or ³⁄₁₆-inch holes near each stud for #8 or #9 steel tie wires. Then have a helper insert a wooden *spreader* at each 2×4 as shown while you twist the wire to secure the walers.

3 Complete the wall form by shoring up the end caps with 2×4s and, after checking for level and plumb, by anchoring it with 2×4 stakes and braces every 6 to 8 feet.

Column, Pedestal, and Pad Forms

When compared with most other forming situations, building forms for columns and pads is a breeze. But don't underestimate their importance or their purpose, which is to provide support for the object that will rest on the concrete.

As a general rule, you should extend the form to below the frost line if the structure to be supported will be tied to an existing structure, as would be the case with an attached deck. This prevents structural damage that could occur as a result of frost heave. In all other cases, you should be safe excavating to a depth of 18 to 24 inches below ground level.

1 As you can see from our first example, grade-level columns are simply holes in the ground. No fancy framing, no fuss. Just determine the correct location and excavate, using either a clam-shell or auger-type digger.

If your plans call for continuing the column above grade, you have a couple of options, also shown. With the first, you dig an oversized hole, then insert a commercially available waxed tube and brace it as shown. You can cut the length you want easily with a handsaw. After the concrete has cured, simply peel off the form.

If you would rather, construct a trapezoidal-shaped ¾-inch plywood form, position it over a previously dug hole, and secure it there with bracing.

2 Get air conditioners, transformers, and the like off the ground with a concrete pad formed as shown here. The holes you see serve to stabilize the pad. (Note: Follow manufacturer's instructions as to the minimum slope required for proper operation of air conditioners or heat pump compressor units.)

Nail tube to braces at desired height

Getting Ready for a Pour

Once the forms are securely in place and you have double-checked them for plumb and level, what comes next depends on the project at hand. The sketches that follow show how to deal with several contingencies.

(To prevent the forms from absorbing too much water from the concrete and to facilitate form removal, you should oil any surface that will make contact with the concrete. This is an excellent time to take care of this important detail.)

1 Fill in any low areas within the forms with gravel or sand, then tamp it (here, a 2×4 on edge is being used) to pack the new material in place. Now take a look at the outer surface of the forms. If you spot any gaps between grade and the bottom of the forms, backfill the voids with dirt or sand. Otherwise, some of the concrete will leak out and hurt the looks of the finished product.

2 To conserve concrete on massive projects such as steps, fill some of the space with large stones, chunks of broken concrete, or other filler material. One way to do this job is to lay a low U-shaped stone wall about 6 inches away from the inside of the sidewall forms and to within about 6 inches of the top riser form. Fill it with rubble or well-compacted soil.

3 Unroll and place welded wire mesh so its edges reach within an inch or two of the edge of the finished slab. If you must use more than one piece of mesh to cover an area (it comes in 5-foot-wide rolls),

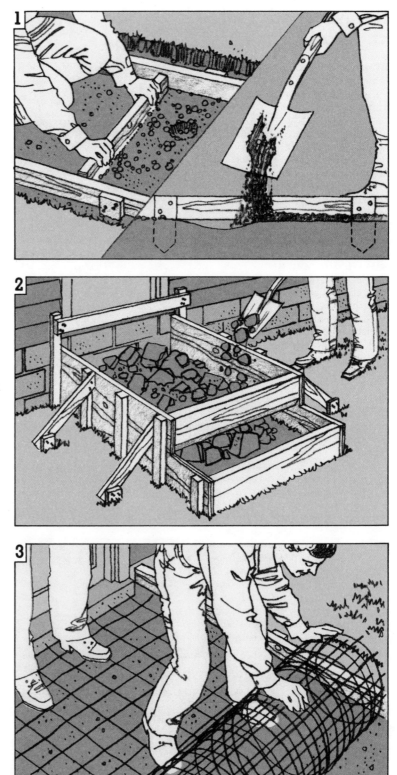

overlap the sections by one square and join the pieces together at several points with wire to keep them from separating when wet concrete covers them. Placing half bricks or rocks 2 or 3 feet apart under the mesh will lift it off the base so it will remain in the middle of the finished slab.

4 To help fortify footings, place reinforcing rods one-third of the way up from the base of the form. Overlap the rod ends by at least 12 inches and tie them tightly together.

5 To prevent flights of concrete steps from drifting or sagging, attach them solidly to the building foundation with reinforcing rod. With *existing poured concrete walls,* create that durable bond by drilling holes at a slight downward angle at least 4 inches deep and drive a length of rod into each. If the rods fit too loosely, grout them in place and bend into final position. Fashion a gridwork by tying intersecting rods to those tied to the wall.

For anchoring steps to an *existing concrete block foundation,* chisel a hole just big enough to stuff wads of paper inside. (This paper prevents concrete from flowing down through the opening at the base of the block.) Fill the whole cavity with concrete, insert the bent rod, and trowel extra concrete around the sides of the rod to ensure a tight fit.

6 Whenever the new concrete will butt up against an existing structure, separate the two with expansion joint material. The gap created allows for expansion of both surfaces. Temporarily position it with a brick or block of wood, which you'll remove while pouring the concrete. Or, you can nail the joint material in place with concrete nails.

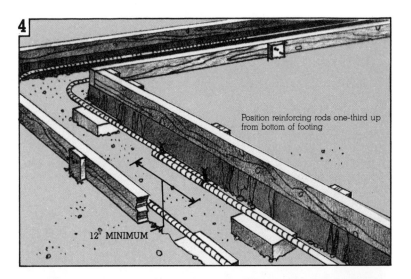

Position reinforcing rods one-third up from bottom of footing

12" MINIMUM

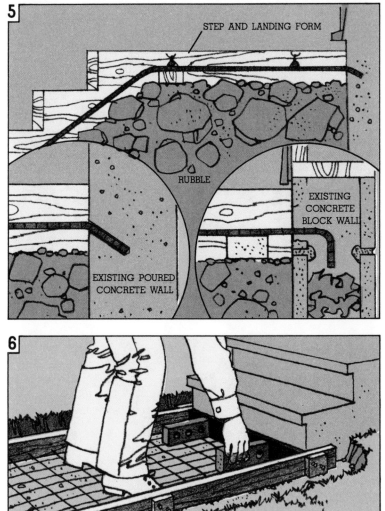

STEP AND LANDING FORM

RUBBLE

EXISTING POURED CONCRETE WALL

EXISTING CONCRETE BLOCK WALL

Estimating Your Concrete Needs

You needn't be a mathematician to figure out how much concrete a job will take. It all boils down to calculating the volume of your form.

If your project is square or rectangular, multiply width by length by height to get volume. For circles or cylindrical forms, multiply the square of the radius by 3.1416 by the height. These calculations will tell you how many cubic feet you're dealing with. To determine the cubic yards of concrete you'll need, simply divide by 27.

For quicker estimates, use the "Concrete Estimator" on this page. And to prevent running short of concrete, increase your estimate by 10 percent.

If you're mixing your own concrete, see the "Concrete Mixing Proportions" chart opposite to figure the amounts of ingredients you'll need.

For additional planning pointers, see pages 10–13.

1 To estimate your concrete needs for this slab, multiply width times length times depth: $10' \times 12' \times .33'$ (4" = .33') = 39.6, which rounds off to 40 cubic feet. Divide this by 27, add 10 percent for waste, and you'll find that you need 1⅔ cubic yards.

2 Divide flights of stairs or other odd-shaped jobs into easy-to-figure sections for estimating purposes. Then add the results to get total volume. In this example, the steps have 8-inch (.66 foot) risers and 10-inch (.83 foot) treads and the landing is 4×4 feet. Calculate length × width × depth = volume. For the first step, $4' \times .83' \times .66' = 2.19$ cubic feet. For the second step, $4' \times .83' \times 1.32' = 4.38$ cubic feet. For the third step $4' \times .83' \times 1.98' = 6.57$ cubic feet. For the landing, $4' \times 4' \times 2.64' = 40.24$ cubic feet. The total volume thus far is 53.38 cubic feet.

Divide 53 cubic feet by 27 and you get a bit less than 2 cubic yards. Now figure the footing's volume and divide the result by 27. Add this cubic yardage to that for the steps and landing, and add an additional 10 percent to cover waste.

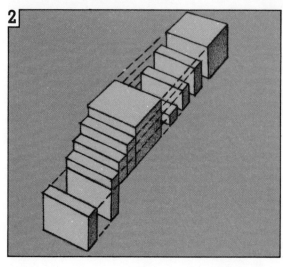

CONCRETE ESTIMATOR
In cubic feet and (cubic yards)

Thickness	Surface area of job in square feet				
	20	50	100	200	500
4 inches	6.7 (.2)	16.7 (.6)	33.3 (1.2)	66.7 (2.5)	166.7 (6.2)
6 inches	10 (.4)	25 (.9)	50 (1.9)	100 (3.7)	250 (9.3)
8 inches	13.3 (.5)	33.3 (1.3)	66.6 (2.5)	133.3 (5.9)	333.3 (12.5)

Mixing or Ordering Concrete

You can buy concrete three different ways. For small jobs, such as setting a few fence posts, a premixed concrete is your most practical choice. It usually comes in bags weighing up to 80 pounds that contain cement, sand, and aggregate mixed together. Each sack contains enough mixture to make anywhere from one-third to two-thirds of a cubic foot of concrete.

For jobs that take one-half cubic yard of concrete or more, either buy the cement, sand, and aggregate and mix them yourself, or order ready-mix concrete. Mixing it yourself (using 10 shovelfuls of portland cement, 22 shovelfuls of sand, and 30 shovelfuls of aggregate for each cubic foot) allows you to make a small batch at a time, and place and finish one section before repeating the process.

If you're time and energy conscious, order ready-mix concrete. Not only will you get a more reliable mix, but you'll also avoid hauling and mixing messes that occur when working with ingredients. **NOTE:** The ideal temperature for working concrete is 70 degrees F. Warmer or colder temperatures hasten or retard the rate of setting and curing. Avoid working in severe cold to prevent the concrete from freezing, which can ruin it.

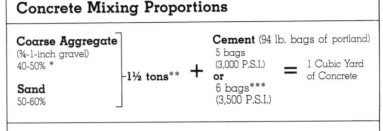

Concrete Mixing Proportions

Coarse Aggregate (¾-1-inch gravel) 40-50% *			Cement (94 lb. bags of portland)
Sand 50-60%	]1½ tons** +	=	5 bags (3,000 P.S.I.) or 6 bags*** (3,500 P.S.I.)	1 Cubic Yard of Concrete

* The more aggregate, the stronger the mix.
** These ingredients can be purchased separately (in the proportions described here) or together in a mixture commonly known as con-mix.
*** For projects requiring extra lateral strength such as a driveway.

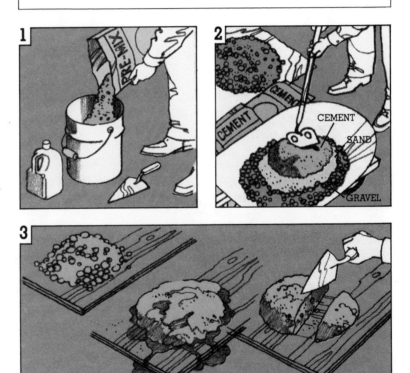

1 With a pre-mix, you simply add water, stir, and use. Mix the contents until all traces of powder are gone.

2 Layering your gravel, sand, and cement as shown helps achieve a consistent mix in the shortest time. Hollow out a small hole in the middle of the dry material, pour in a little water, and hoe edges into the puddle. Add water judiciously as you need it to achieve the mix you want.

3 Controlling the wetness or *slump* of your concrete is just a matter of watching it as you mix in water. The leftmost example is a mix that's too dry to handle easily and to cure to full strength. In the middle is a mix that's too soupy. (Overwatering is probably the most common mixing mistake.)

The rightmost example shows a proper mix that holds most of its shape when sliced with a trowel but is still soft enough to pour and shape as you wish.

32

Placing Concrete

Once the concrete starts to flow, it's too late to alter the forms, to run for tools, or to look for more help. So, before you start mixing or before the ready-mix truck arrives, be extra sure you're ready.

For placing an average job, you'll need a shovel, a wheelbarrow, a rake, a straightedge for screeding, a bull float or darby, a wood or metal float, a steel trowel, a jointer, an edger, a broom, and a water hose. You'll use some of these for finishing the concrete.

Unless the compacted base is still damp from an earlier wetting, sprinkle it and the forms with water. This will help prevent a too-rapid moisture loss from the concrete you place. Sprinkling the site just beforehand is especially important if you're placing concrete on a warm, windy day.

On all but the smallest projects, you should have two or more strong (if not entirely eager) helpers. And if you're undertaking a large job such as a driveway or a sizable patio, have enough help on hand so some of them can rest occasionally while the others work.

1 To save time and muscle strain, mix your materials as near the job site as you can or have the ready-mix truck park as close as is safe. (Be advised, however, that the weight of a full concrete truck is immense.)

Wet concrete weighs 150 pounds per cubic foot, so when wheeling it, keep loads small enough to

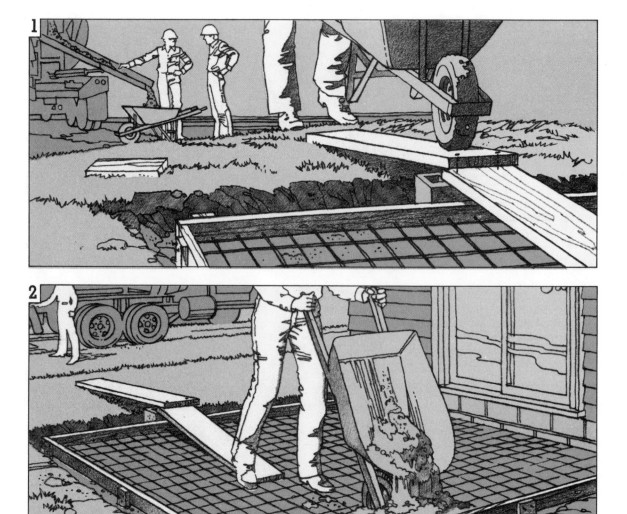

handle easily. To cross soft soil or lawns with a loaded wheelbarrow, lay a walkway of 2×10 or 2×12 planks and build ramps into the forms so you don't disturb them. Use two or more wheelbarrows for larger projects, especially if you're moving concrete more than a few feet (this speeds the entire process).

2 Start placing the fresh material in the farthest corner, dumping it in mounds that reach ½ inch or so above the top of the form.

3 As you pour, tamp the concrete to fill any air pockets, especially in the corners and all along the perimeter.

4 While pouring, use a hook, a claw hammer, or a notched spade to pull the wire mesh up into the concrete (see page 28 for details on placing the mesh). For top strength, keep the mesh near the center of the slab's thickness.

5 Begin screeding as soon as you've filled the first 3 or 4 feet of the length of the form. This will show you whether you're putting in enough material.

Keep both ends pressed down on the top of the form while moving the screed back and forth in a sawing motion and drawing it horizontally into the unleveled concrete. If hollow

spots appear in the screeded surface, shovel in more material. Then *strike off* (level off) the newly filled areas, working toward the unscreeded ones.

6 After you finish the screeding operation, use a darby or a bull float to further smooth the surface and to imbed the aggregate below the surface. Now take a break, and review the finishing and curing procedures on the following six pages.

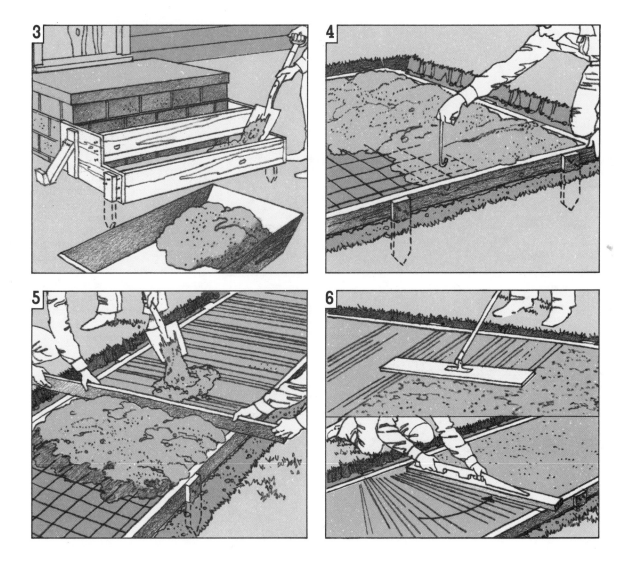

Finishing Concrete

Any concrete finisher will tell you that the secret to the finishing operation is in the timing. Start too soon and you'll weaken the surface; too late and, well, forget it.

The waiting period between placing and finishing varies depending on wind, temperature, humidity, and type of concrete you use. To be safe, follow this rule: *Start finishing concrete when water sheen is gone from surface, and concrete will carry your foot pressure without sinking more than ¼ inch.*

Edging and Jointing

1 Edging makes neat, rounded edges that resist damage and look attractive. It also compacts and hardens the concrete along the form, where floats and trowels don't reach adequately.

Hold the edger flat on the concrete surface with the front tilted up slightly when moving ahead with it. Then raise the rear slightly when drawing it backwards. Use short back-and-forth strokes to start shaping the edge and to work larger pieces of gravel deeper into the concrete. Then use long, smooth strokes, holding the tool level. Be careful not to press too hard; this may leave low spots that are difficult to remove. Repeat the edging process after each other finishing task.

2 Jointing results in a series of grooves whose purpose it is to prevent random cracking that would ruin the concrete's appearance. You can fashion these so-called *control joints* by finishing them into the wet concrete with a jointer, or by sawing grooves in the surface after the concrete has hardened. To work well, a control joint should have a depth equal to one-fourth of the vertical thickness of the slab (which typically is ¾ inch).

Space joints in driveways and sidewalks at distances equal to the slab width. With patios or drives wider than 10 feet, also run a joint down the center. Intersecting joints should be roughly square. In general, the smaller the panel, the less likely it is to crack later because of expansion.

Work the jointing tool as you would an edger, using a straight 2×4 as a guide. If you decide to cut in control joints afterwards with a circular saw, outfit the saw with a masonry cutting blade and saw grooves to a depth of one-fourth the thickness of the slab. Do this as soon as the concrete is hard enough not to tear on impact of the blade—normally a few hours after the concrete hardens. To avoid cracking, complete sawing within 12 hours after the concrete sets.

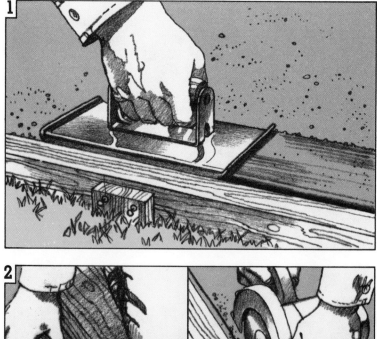

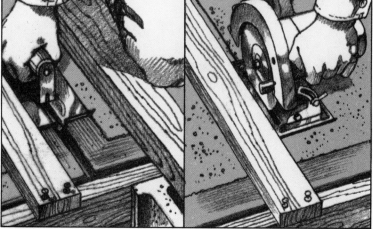

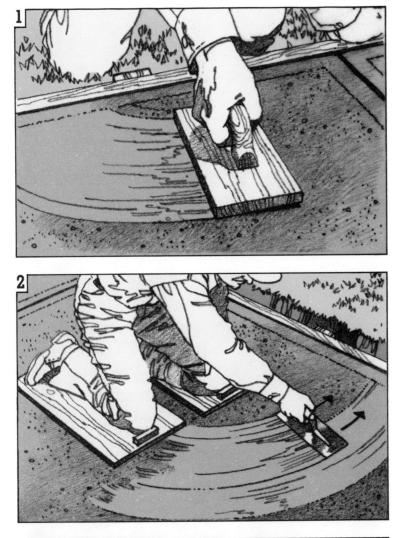

Floating, Troweling, and Brooming

1 Floating does three things: It pushes larger pieces of aggregate deeper into the concrete; it smooths the surface; and it draws a mixture of sand and cement to the surface, making further finishing possible. The finish a float leaves depends on the material it's made of. For example, a wooden float produces a coarse-textured surface, and a magnesium float makes a smoother one.

If water begins to surface when you begin the floating operation, STOP, wait a while, and try it again. Hold the float nearly flat and sweep it in wide arcs to fill low spots, flatten lumps, and even off ridges. Be sure to smooth up marks left by edging and jointing.

2 To achieve a smooth, dense surface, switch to a steel trowel after floating. Hold the trowel blade nearly flat against the surface. Overlap each pass by one-half of the tool's length so you trowel all of the surface twice in the first operation. For an even smoother surface, trowel again a second or even a third time after intermittent waits. The final time, the trowel should make a ringing sound as you move it over the concrete, with the leading edge raised slightly.

3 For a nice-looking, slip-resistant surface on steps, walks, and drives, pull a damp broom across the surface of just-troweled concrete. For a fine texture, use a soft bristled broom; for a coarser one, use a stiffer bristle. You get best results with a special broom made for this job.

Custom Finishes

On the previous two pages we covered the steps necessary to finish concrete and to apply a broomed or smooth finish. But you may elect to give your project a special twist via an out-of-the-ordinary look. Here are some interesting possibilities to consider.

1 Maybe you want to add a splash of color to your project. If so, adding color pigment to the concrete is an obvious and not-overly expensive way to get the job done (unless your concrete supplier adds the color). For large jobs, such as patios or sidewalks, professional concrete finishers usually shake about half of the required amount of powdered concrete color over the area to be tinted right after edging and floating.

As soon as the powder becomes wet, edge and float the surface again to spread the pigment evenly. Now shake on the remainder of the powder and repeat the edging and floating processes.

For smaller jobs, you can mix the pigment in right along with the other ingredients. Keep in mind, however, that coloring the entire mix requires much more powder. Note, too, that achieving consistency from batch to batch requires careful measurement of ingredients.

2 For an exposed aggregate finish, place the concrete and screed it to about ½ inch below the top of the forms to make room for the pebbles you will embed in the surface.

After the water evaporates from the surface, sprinkle stones so they cover the concrete in a uniform layer. With a float, carefully "mush" the aggregate into the concrete until the surface looks about like a normal slab that has just been floated. If you are finishing a large slab this way, you may want to apply a curing compound first to slow the curing action of the concrete and extend your working time.

When the concrete is firm enough to support a person on kneeboards

without leaving marks (one to two hours after it's poured) start brushing away excess concrete with a stiff nylon-bristle broom. After the initial brushing, spray a fine mist over the surface. Continue sweeping and misting until the runoff water is clear and the top one-third to one-half of the aggregate shows no cement.

3 If a travertine finish appeals to you, start by spattering on freshly leveled concrete a coat of colored grout that's the consistency of thick paint. After the mortar stiffens slightly, use a steel trowel to smooth high spots. (Note: Freezing moisture will wreck this finish, so use it only if you live in a mild climate or where it will be protected from the weather.)

4 To make patterns that resemble flagstones or a geometric shape, score the concrete soon after bull-floating or darbying and after any bleed water evaporates. A joint strike (see page 7) works well for this. Go over "joints" each time you do other finishing operations.

5 Combining custom finishes can produce pleasing effects, too. Here, sections finished in a wavy broomed pattern adjoin sections of exposed aggregate.

Curing Concrete

How important is this last stage of any concrete project? Very! Why? Because *hydration,* the chemical process that hardens cement, will stop after the concrete first sets unless you keep it moist and fairly warm. And without adequate hydration, the concrete ends up weaker and softer than it should be. Concrete actually continues to get stronger for years, but the first week after you pour is the critical period.

You can successfully cure concrete several ways: by providing wet coverings, sprinkling, or ponding; by covering the newly laid material with sheets of plastic or waterproof paper; by covering the concrete with straw; or by sealing the surface with a curing compound. Any of these measures will help keep moisture from leaching out of the concrete prematurely. Which way is best? If you're willing to spend the time keeping the surface wet constantly, sprinkling is the way to go.

NOTE: Extreme temperatures cause problems in placing, finishing, and curing concrete. If the temperature soars above 90 degrees F, for example, the concrete will set faster, so you must finish it more quickly. Too, there's a risk of fine "chicken wire" cracks appearing in the surface because of too-rapid drying.

Ways to avoid these headaches include working in the cooler early-morning or late-evening hours, soaking the subgrade and forms before placing the concrete, reducing finishing time by having plenty of helpers, covering the wet concrete with a tarp or plastic sheet between placing and finishing, and putting up windbreaks or shades. You can also fog an extremely fine mist over the wet concrete to stop evaporation. Careful, however; you don't want to add water to the mix itself. Cold weather, too, presents difficulties you need to be aware of. If newly placed concrete freezes before curing, its structural integrity is questionable, again because the water needed for hydration hasn't been allowed to perform its important function.

So if you know of an approaching front that's expected to bring with it extremely hot or cold weather, delay your plans for a while. And if you get caught off guard, protect the project with straw as shown in sketch 4.

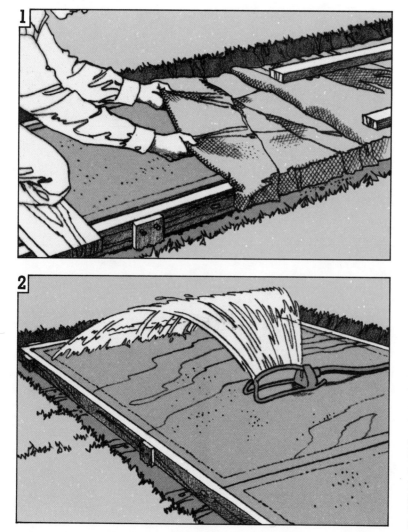

1 Burlap or old blankets, kept wet by frequent watering, are the most common covering. Be sure to use clean fabric so the fibers won't stain or chemically harm the concrete. As soon as the concrete is hard enough to resist surface damage, cover it with burlap or similar material weighted down by scrap lumber. Sprinkle it often enough to keep the covering wet for the whole curing period—at around 70°F. Five days should be sufficient.

If yours is a vertical pour, leaving the forms in place for this duration serves as a more practical way of providing curing protection.

2 Constant wetting (especially during daylight hours) with a lawn sprinkler also will provide the moisture curing concrete needs. Make sure the "set" of the new work has advanced enough so the streams of water don't damage the new concrete surface (which could cause spalling later).

3 Because they offer a convenient (one-time application) way to protect curing concrete, pigmented curing compounds are very popular, especially among professional concrete finishers. Nevertheless, these compounds don't work quite as well as moist-curing. If you go this route, apply the compound while the finished concrete is still damp, but not actually wet. For thorough coverage, spray a second coat, working at a 90-degree angle to the first one.

4 One effective way to cope with cold weather if you are pouring concrete is to use heated water for mixing it. This way, the concrete goes into the forms warm. As it sets up, concrete produces some additional heat. But if cold weather strikes, you'll also need to spread 6 to 12 inches of dry straw or hay over the concrete, then cover the straw with canvas or polyethylene film. Be especially careful to cover the edges and corners of slabs, the parts most likely to freeze.

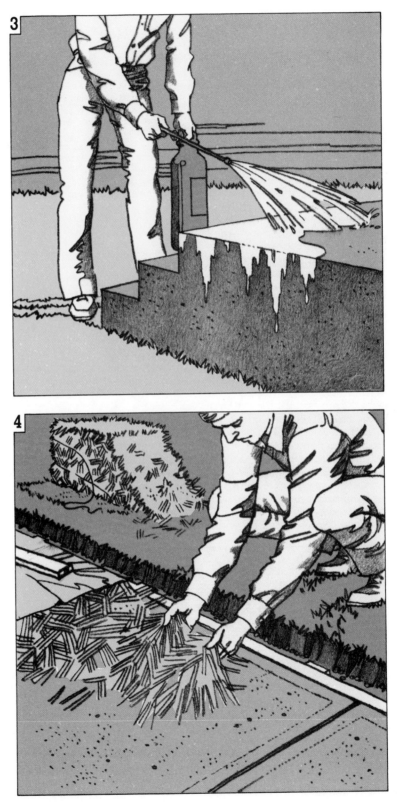

Repairing Concrete Surface

When you have problems with concrete, whether it be *spalling* (flaking) or a *structural defect* such as a crack or break because of settling, you can bet that one of two things has happened. Either the freeze-thaw phenomenon has caused a fracture, or the concrete was placed incorrectly. Regardless of the cause, however, you're left with an unsightly, and, in some cases, potentially damaging, gap in the structure's surface.

How you repair the defect depends, obviously, on the situation. You can renew spalled surfaces or those with hairline cracks and small breaks with the patch materials discussed in the Concrete Patch Selector (right). To deal effectively with settled or badly cracked surfaces, however, you'll need to remove the offending section and place new concrete (see below).

Concrete Patch Selector		
Type	Uses	Description/Mixing Instructions
Latex patch **Vinyl patch** **Epoxy patch**	General-purpose repair jobs such as filling hairline cracks, small breaks and patches, and tuckpointing. Latex and epoxy are best for patio surfaces.	Sold in powdered form with or without a liquid binder. Mix with the appropriate binder, usually a sticky white liquid, to a whipped-cream consistency.
Hydraulic cement	To plug water leaks in masonry walls and floor surfaces. This fast-drying formulation allows you to make the repair even while the water is leaking in.	Available in powdered form. Mix a small amount with water or a commercial binder, then work quickly.
Dry premixed concrete	Wherever whole sections of concrete are being repaired.	Mix this product with water. The thicker the consistency the faster it will set up.
Note: Mixing a synthetic binder with the patch material strengthens the repair and enables the patch to adhere more securely to the old surface.		

1 To repair a spalled area, start by using a cold chisel (held at an angle) and a 2-pound sledge to remove the top ½ to 1 inch of the surface. Undercut the edges to aid in locking together the new and existing concrete.

CAUTION: Chips will fly as you chisel the concrete, so wear safety glasses or other eye protection as you work. Wear work gloves, too.

2 Brush out large particles and flush away dust and any remaining pieces of concrete with a strong stream of water. Some moisture will soak into the old concrete, but that's good. It should be damp when you apply patching material. Sponge out any standing water, however.

3 Trowel in slightly more patch material than you need to fill the cleaned-out hole; then, for small patches, screed off the excess with a trowel. If you're patching an area wider than the length of your longest trowel or float, use an adequate length of straight lumber as a screed. Finishing the patch, following the steps presented on pages 34–35.

4 To replace a severely cracked or broken section of concrete, start by scoring the bottom of the control joints (grooves) that separate the damaged section from adjacent ones. Use a brickset or stone mason's chisel and a 2-pound sledge with enough force to leave a line of white marks. (The purpose of scoring is to make a row of hairline cracks that will prevent random cracking beyond the expansion joint as you work.)

Break out pieces by hammering at the same spot with an 8- or 10-pound mall until the concrete fractures. Once you have removed enough to be able to pry up large pieces of the slab, do so and place a piece of rubble underneath, as shown in the detail. This makes it easier to break off additional chunks. Carefully chip away final areas along the control joints. Clear away the waste.

5 Install forms and expansion material as shown here, spread sand or crushed rock to fill in low spots in the subgrade, and lay in wire mesh to strengthen the new section. Place a normal concrete mix in the form and finish it as discussed on pages 34–35, while trying to duplicate the finish texture of the existing concrete as closely as possible.

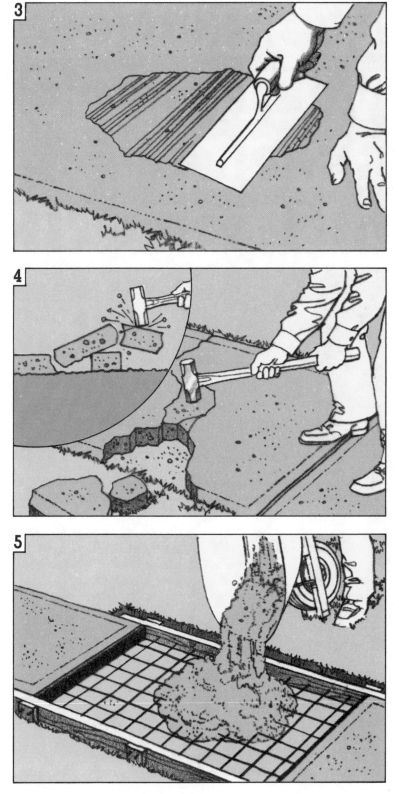

WORKING WITH BLOCK, BRICK, AND STONE

Most people admire the skill needed to build a handsome fireplace facing, a multi-story structure wrapped in a beautiful brick veneer, a stone walkway, and almost any other masonry project you can mention.

But you probably couldn't find one person in a thousand who thinks he could handle a masonry project himself. For some reason, most people assume that masonry is one skill only a veteran craftsman can handle. But that's simply not the case. This second section, we hope, will help dispel any misunderstandings you may have about masonry.

For success with masonry projects, remember these three secrets: adequate preparation, knowledge of a few basic rules, and application of some age-old techniques. On the following pages we cover all three, with special emphasis on the techniques, which we cover step by step. We urge you to read the entire section before beginning any project. By doing so, you'll learn about certain situations and problems that can arise—and how to best handle them.

Determining Your Material Needs

No matter what type of masonry project you undertake, your No. 1 priority is determining the *area* the material —be it brick, block, or stone —will cover. This requires that you recall a bit of high school geometry and brush up on the formula: area = length × width.

Once you calculate the project's area (you'll get so many square feet and don't forget to subtract from this the area of any openings), refer to the chart below. Based on nominal dimensions where applicable, it will help you translate square feet into so many bricks or blocks or into so many tons of stone.

You can also use the chart to help figure your mortar needs. It shows how many bags of *pre-mixed mortar* (and how many bags of *masonry cement and sand*) you'll need for so many bricks and blocks.

(**NOTE: After determining your needs, add 5 percent to the total to allow for waste.**)

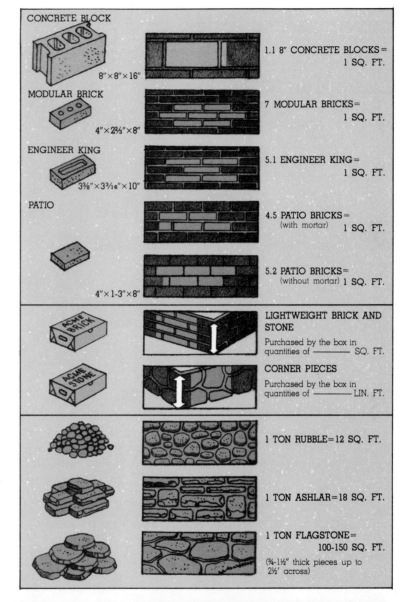

CONCRETE BLOCK
8"×8"×16"
1.1 8" CONCRETE BLOCKS = 1 SQ. FT.

MODULAR BRICK
4"×2⅔"×8"
7 MODULAR BRICKS = 1 SQ. FT.

ENGINEER KING
3⅜"×3³⁄₁₆"×10"
5.1 ENGINEER KING = 1 SQ. FT.

PATIO
4.5 PATIO BRICKS = (with mortar) 1 SQ. FT.

4"×1-3"×8"
5.2 PATIO BRICKS = (without mortar) 1 SQ. FT.

LIGHTWEIGHT BRICK AND STONE
Purchased by the box in quantities of ———— SQ. FT.

CORNER PIECES
Purchased by the box in quantities of ———— LIN. FT.

1 TON RUBBLE = 12 SQ. FT.

1 TON ASHLAR = 18 SQ. FT.

1 TON FLAGSTONE = 100-150 SQ. FT.
(¾-1½" thick pieces up to 2½' across)

1 70-LB. BAG
PRE-MIXED MORTAR = 30 PATIO BRICKS
(Bag sizes vary, depending on manufacturer)   40 MODULAR BRICKS
12 BLOCKS

1 TON
MASONRY SAND = 800 PATIO BRICKS
1000 MODULAR BRICKS
300 BLOCKS

1 70-LB. BAG
MASONRY CEMENT = 80 PATIO BRICKS
(Plus masonry sand)   100 MODULAR BRICKS
30 BLOCKS

1.5 TONS OF ¾" GRAVEL = 1 CU. YD.
(For drainage)

Note: Calculate *volume* in cubic feet, then divide by 27 to figure *cubic yards.*

44

Laying Block, Brick, and Stone Without Mortar

Patios and Walks

If the idea of building a good-looking, durable patio or walk simply by laying masonry units side by side on a bed of level sand sounds impossible, you're in for a pleasant surprise. It's a snap!

And you needn't lay a single trowel-full of mortar. Just brushing or sweeping loose sand into the joints forms a surprisingly solid surface. And talk about quick. You won't believe it.

The resulting surface is called *flexible paving*. Even

in colder climates where frost heave will cause the surface to rise and fall, this type of surface holds up surprisingly well year after year.

The choice of patterns you can use for laying block, brick, and stone is practically unlimited. Just a few of the possibilities appear here. Many of the old standards are quite appealing, but you may opt for a design uniquely yours. Whichever way you decide to go, make a dry run—lay a sample section—to see how well you like it. Leave a ⅛- to ¼-inch space between blocks or bricks and a ½-inch space between flagstones.

You'll want to lay some type of border around the patio or walk, too. Borders serve two purposes— one ornamental, the other functional. Not only do they add a distinctive touch to the project, they also prevent the units from shifting or working loose. The sketch at left features four attractive border options to consider. These include setting bricks on end, diagonally on end, or butted end-to-end around the perimeter. The fourth possibility uses redwood or cedar 2×4s as divider strips and borders. You'll want to stake a wooden border every 2½ feet to anchor the members.

1 Prepare the site by staking out the perimeters, removing sod, and excavating (see pages 15–16 for details). When excavating, keep in mind that you want the paving material to be 1 inch above grade when you're through. Be sure to allow for a 1- to 2-inch bed of sand, or where moisture is a problem, a 2-inch bed of crushed rock

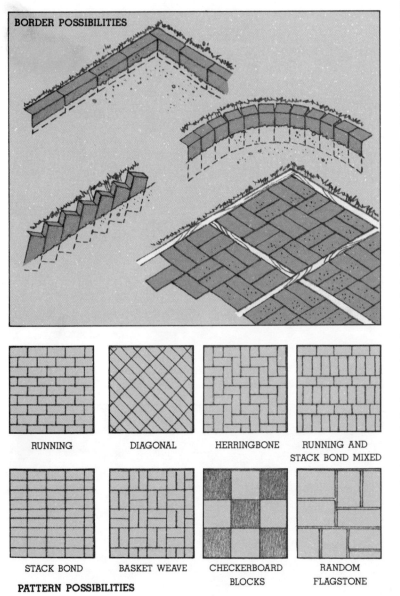

BORDER POSSIBILITIES

RUNNING   DIAGONAL   HERRINGBONE   RUNNING AND STACK BOND MIXED

STACK BOND   BASKET WEAVE   CHECKERBOARD BLOCKS   RANDOM FLAGSTONE

PATTERN POSSIBILITIES

plus 1 to 2 inches of sand to provide the necessary drainage. When trenching for the border, remove just enough soil so the border rests on firm ground.

2 Stretch a mason's line to serve as a guide for the height and alignment of the border. Place a small amount of sand in the bottom of the trench and tap the bricks with the handle of your trowel to achieve the height you want. Finally, fill around the sides of each brick with sand as you work along the trench. For an even longer-lasting border, fill the trench partway with mortar, then place the bricks firmly in it as described above.

3 To provide a level bed for the main surface, draw a straightedge across the sand bed to smooth it. A straight 2×4 with a length of 1×4 or 1×6 nailed to it is useful for striking off the sand at the right level. If your striker is too short to reach across the patio, place stakes and a 2×4 rail at the proper height and use as shown. *(continued)*

Patios and Walks

(continued)

4 Cover the leveled sand base with 15-pound asphalt-saturated felt to prevent weeds from sprouting up between the masonry units and to keep moisture from staining them. (Sheets of dark polyethylene plastic also will work for this purpose.) Overlap sheets 2 inches.

5 For a patio or walk to effectively and quickly shed moisture, its surface must have the correct elevation and adequate slope. For patios this means the surface should be at least 1 inch above grade at the lowest point, and have a slope of ¼ inch per foot of length. Walks, too, should be slightly above grade and slope ¼ inch per foot across their width.

    As you place the units, use a level to check slope and as a straightedge to make sure the bricks are all the same height. If a brick, block, or stone is too low, pick it up, trowel in more sand onto its base, and tap the unit into place until it rests at the desired height. Using a taut line as a guide will help you lay the units in neat, straight courses at the proper elevation.

6 To successfully cut a brick, stone, or concrete block, first score it along the line of the cut. Mark the cutoff line with a pencil, then place the unit on a bed of sand or loose soil several inches deep. Hold a brick set on the cutoff line and tap it with a hammer, using enough force to leave a mark. On concrete blocks and stones, where the line of the cut is longer than the width of the brick set, work along until you have scored the whole line.

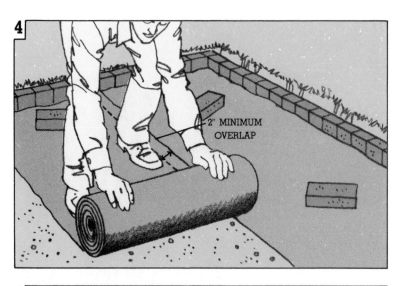

2" MINIMUM OVERLAP

To make the final cut, hold the brick set perpendicular to the surface with the bevel of the blade away from the side of the finished cut. Rap it sharply and the unit should break on the line or near it. If it doesn't break cleanly, you'll have to chip off the rough edges, with short, brisk strokes using the chisel end of a mason's hammer.

After scoring a larger stone on both sides of the cut line, place one end of it on a solid support such as a wood block. Tap the unsupported end with a hammer and the unit should break along the score line. With practice, you'll be able to cut bricks, blocks, and stone with expert control.

7 Spread a thin layer of sand over the surface of the newly laid units and gently sweep back and forth with a broom so the sand fills the joints. Be especially careful at first not to dislodge the units.

8 Once the joints are full of sand, hose off the entire surface with a fairly fine spray to wash away the remaining fine particles. The moisture will help compact the sand in joints. If it settles too much, sweep in more sand and spray again.

NOTE: If desired, you can add significantly to the permanence of flexible pavement by using a dry mortar mix in place of plain sand in the joints. For this method, leave ½-inch spaces between block and brick. Sweep the dry mortar mix into the joints, remove any excess, and sprinkle with water until the mix is wet. Repeat sprinkling twice at 15-minute intervals to ensure that you have enough water in the mortar. It will harden within a few hours and cure in one week. See pages 52–53 for more details on laying patios and walks with a mortar mix.

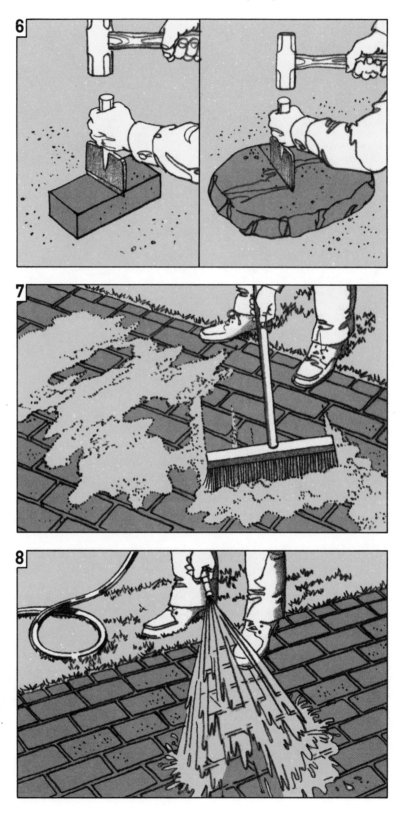

Building Dry Walls

Building a wall of natural stone is hard work (and fairly expensive), but the results usually are strikingly beautiful as well as functional. Here we show how to erect both *freestanding* and *retaining* walls without mortar. These so-called *dry walls* rely on gravity rather than mortar to hold them together.

When planning your project, keep in mind the following two guidelines: First, build the wall no higher than 3 feet tall; for retaining walls, if the top grade you're building to is higher than 3 feet, build in step-like terraces. Second, make sure the wall tapers in slightly from the base of the top of the wall. For information on how to estimate the amount of stone needed for your project, see page 43.

NOTE: When lifting natural stone, do so with your legs, keeping your back straight and rigid to avoid straining it. With larger stones, get someone to help you.

Freestanding Walls

1 Begin by driving stakes and stretching string lines to establish the edges of the bottom of the wall. Remove sod and 2 or 3 inches of soil to provide a smooth, level base for the first course and allow its top to rest slightly above grade. Begin setting the stones, making sure you set those in each course perpendicular to those in the adjoining course. This helps tie the courses together.

2 Fill any voids in the center of the wall with small stones. And plug vertical gaps between stones by tapping small pieces of stone *(chinks)* into place as shown. With

TIE STONE

a brick hammer and some practice, you should be able to make wedges of the size you want.

3 Here's what a cross-section of your wall should look like near its completion. Notice how the stones form bonds both across the width and along the length of the wall. For added strength, select large, flat stones for the top course. (Some stone masons spread mortar over the stones in the next-to-top course, then set the top course of stones into the mortar. This helps seal the top of the wall from moisture, which otherwise may weaken the wall.)

Retaining Walls

In many localities building codes apply to stone retaining walls. So, before you begin, check with local authorities to see whether any such rules apply to your project. If they do, follow them wherever they vary from these instructions.

1 After staking and laying out the front perimeter, excavate a base for the bottom course of stone. If any stone is loose when you first place it, trowel some loose dirt under and around it to stabilize it.

2 To cut large stones to fit, start by scoring all the way around the stone along the line of the intended break (be sure to wear gloves and safety goggles). Then place the stone on a solid support along the score line and tap the unsupported part until it breaks off.

3 Slope the face of the retaining wall 10 to 20 degrees from vertical so it will lean toward the backfill it is to retain. To keep the middle of the wall from bulging, about halfway up install long *tie stones* a maximum of every 4 to 6 feet.

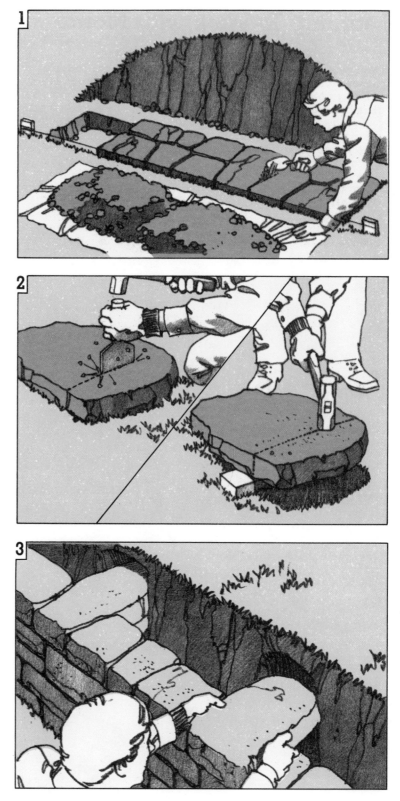

50

Learning To Work with Mortar

If you've ever been around a journeyman mason, you've probably noticed that he takes his mortar very seriously. He may, for example, refuse a batch of mortar as being too soupy, or have the laborer add more water to make the "mud" work better. It's got to be just right!

Why all the fuss? Because a skilled mason knows that the quality of the final product depends largely on the qual-

ity of the mortar. That's a good lesson for you to learn early on, too.

How you go about preparing the mortar for a job depends on the project's size. For small ventures and repair work, it's most convenient to use premixed mortar that already contains sand. However, for larger jobs, you'll save money by buying the sand and masonry cement separately, and mixing them.

For help with estimating the amount of mortar needed for a project, see page 43.

NOTE: Cold weather adversely affects the strength of mortar; so if you expect the temperature to dip below 40°F, hold off on your project and wait for warmer weather.

1 Carefully measure the sand and masonry cement into the container you'll use for mixing. (For small batches, a contractor's wheelbarrow will do, but with larger amounts you'll need a mortarbox.) *Use 3 parts sand to 1 part cement.* (To further strengthen mortar, you can add 1¼ shovelfuls of portland cement per 70-pound bag of masonry cement.) Spread half of the sand first, then the cement, and finally the rest of the sand. (Mix up only as much as you can use in an hour, which will take some experience to determine.)

With a chopping action of the hoe, mix the dry material thoroughly by pulling and pushing it back and forth. When you add water, measure it, too. This usually will amount to 2½ gallons per batch containing 1 bag of cement. This varies, however, depending on the moisture content of the sand. Don't use a hose to run water into the mix—it's too easy to "drown the mortar." If you do accidentally get it too thin, add properly proportioned dry mix to stiffen it.

CAUTION: Because cement is slightly caustic and sand is abrasive, wear gloves and adequate eye protection as you work.

2 For colored mortar, add a suitable mortar pigment to your dry mix. (Be sure to follow label directions carefully as to the maximum

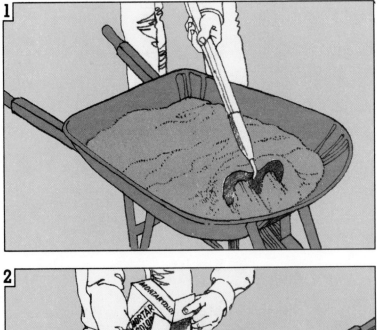

amounts permissible; using too much pigment weakens mortar.) As you add the pigment to the first batch, make a note of the amount you use to obtain the color you want so you'll know how much to add to successive batches for consistent coloring.

NOTE: You *cannot* retemper (remoisten) colored mortar as it begins to set up without altering its color significantly.

3 How can you tell when your mortar is "right" in terms of consistency and adhesiveness? Simply pick up a small amount of it with your trowel, stick it to the trowel with an upward jerk, and turn the trowel upside down. If the mortar adheres to the inverted trowel, it has the consistency you want. If mixed mortar stiffens before you can use it all, you can retemper it once—add water and remix—with no complications (this applies only to non-colored mortar). But if it starts to dry up again, discard it and mix a new batch.

4 It's easier to load a trowel with mortar if you slice off a gob and roll it in a windrow to the edge of the mortarboard. Masons call this process *cupping* the mortar. Once you've made the windrow, scoop it up with a smooth sweep from one end. With an upward jerk of the wrist, stick the mortar to the trowel. Jerking too hard will dislodge it.

For convenience as you work, place your mortarboard or mortar pan a couple of feet away from the area where you're laying bricks or other units. Also support it at a comfortable height so you needn't bend over excessively.

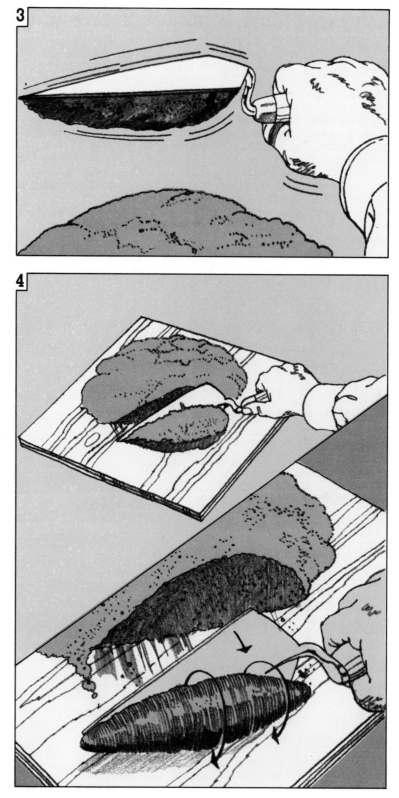

Laying Block, Brick, or Stone with Mortar

Patios and Walks

Paving an existing concrete slab patio or walk with a layer of brick, tile, or flagstone gives a drab area some real pizzazz. The technique is simple, but somewhat exacting: You spread a layer of mortar on the surface of the slab, embed the new paving material in it, and finish by putting mortar between the pieces of paving.

For the new surface to have a long, useful life, the existing slab must be sound and clean. If it's badly cracked, it may be wiser to put a bed of sand on top and lay the new materials without mortar (see pages 44–47) or to tear up the damaged concrete and start from scratch as explained below.

If you have to pour a concrete slab base first, follow the steps on pages 15–17,

20–21, and 28–33. You won't need to finish the concrete beyond screeding. Leaving it rough will provide a perfect surface for the layer of mortar.

It's also important to allow for plenty of drainage by keeping the surface at least 1 inch above grade and allowing a slope of ¼ inch per foot of length.

1 To establish guides for the elevation of the new surface, stake straight 1×4, 1×6, or 1×8 boards around the perimeter of the existing slab or at least along two opposite sides. Allow enough height between the top of the guide boards and the existing slab to accommodate both the layer of mortar and the paving material embedded in it. For example, if you plan to lay 1½-inch flagstones on a 1-inch bed of mortar, the top of the guide should be approximately 2½ inches higher than the slab surface.

Check the slab's surface for proper slope. If it needs correcting, you can do this by adjusting the guide boards and using extra mortar to build up the surface where it's too low. Use a taut line drawn across the guide boards to check the level of the paving units. Before mixing the mortar (for this, see pages 50–51), make a dry run of all the pieces of paving to inspect both their appearance and their fit.

2 Make sure the surface is clean before starting construction. If it has been painted, you should rough up the surface with an

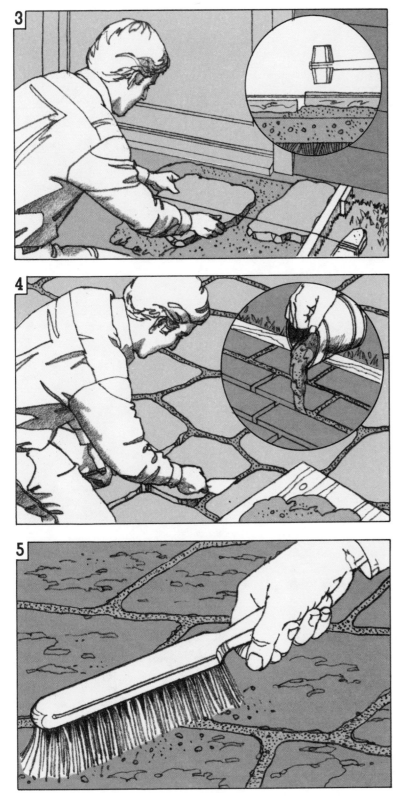

abrasive or have someone sand-blast it. Treat very smooth surfaces with a concrete bonding agent before proceeding further.

Wet the surface just before troweling on an even ½- to 1½-inch layer of mortar. Start in one corner, covering an area of only 4 to 6 square feet. Screed it to a uniform depth with a short piece of 2×4 (see page 33 for more information on screeding).

3 Place the paving units on the mortar, leveling each one as you position it. Remember, too, to level in two directions. If one paver is too high, press it down with the heel of your hand or tap it gently into place with a rubber mallet. Remove those that are too low, trowel in more mortar, and try again.. For small areas or narrow ones such as walks, use a level or a short board laid across the guide boards to check the height of the paving units.

4 Let the mortar holding the paving units set and cure for at least 2 days before you fill the joints. Using a small pointing trowel and the same mortar mix used in the bed, grout the joints. You can fill them so they are flush with the surface or leave them recessed. An alternative way to fill the joints is to pour a slightly thin grout in place with a large tin can that you pinch on one side to form a pouring spout.

5 Brush away excess particles of grout after the joints have hardened for 3 to 4 hours. If cement stains remain after the grout has cured for a week, scrub the surface with water. If necessary, use a 1:15 muriatic acid/water solution to get it clean.

54

Building Block Walls

Watching an experienced crew of blocklayers raise a concrete block wall is like watching the harmonious workings of a clock. You'll see one crew member mixing the mortar, another transporting and dumping it onto a mortarboard, and still another hauling the large, gray building blocks to a location close to the crew leader, who you'll see spreading the mortar with his trowel and adroitly positioning the blocks. All in all, you'll see few, if any, wasted movements.

Obviously, we can't promise that you'll ever be as graceful or speedy as a professional blocklayer. But on this and the following five pages we do point the way—via step-by-step instructions and sketches—to professional-looking block-laying results.

Before getting out your trusty trowel, however, take time to plan the project carefully. Review the information on site preparation (pages 18–19) and on building wall footing forms (pages 20, 24–25). And when determining the dimensions of the wall itself, remember to keep its length in multiples of 8 inches, if possible, to avoid having to cut any blocks. **NOTE: The following instructions, for the most part, cover the steps involved in erecting foundation walls. If your project involves just one wall, you can skip steps 1 and 2. Simply snap a chalk line on your footing between two end points positioned to designate the outside of the block wall, then begin with step 3.**

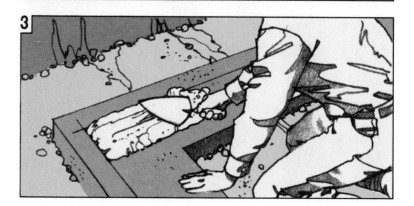

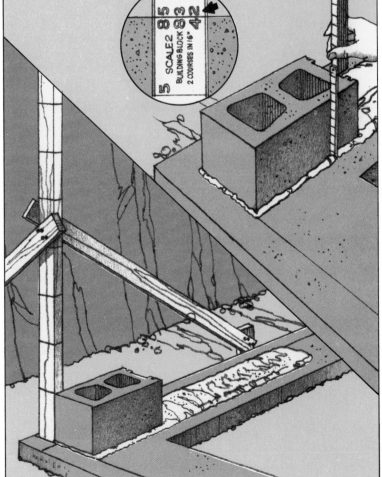

1 Begin by establishing the location of the corners of the wall. To do this, first stretch a mason's line between the batter boards you set up before excavating. (Place the line in the kerf marking the building [wall] line; see pages 18–19 for help with this.) Now, with a helper, dangle a plumb bob from the points at which the mason's lines intersect. (Be careful you don't disturb the alignment of the lines.) Mark the locations on the footings with a pencil.

2 Snap chalk lines between all corner marks, then check the adjoining lines for square, using the 3-4-5 method on page 15.

3 Sweep the footing clean of any dirt or debris, then lay a solid 1-inch bed of mortar for the first course of block, starting at one corner and running the length of 3 or 4 blocks. Make the bed about 1 inch wider than the block you'll be placing. (Be careful not to cover up the chalk line.)

4 Set a corner block carefully into position, *with the smaller holes in its cores facing up and its "ears" pointed toward the opposite corner.* Press it gently into place.

5 Measure to confirm that the joint is exactly ⅜ inch thick. If you use a conventional folding rule or tape measure, the top of the block should be 8 inches above the footing. On a modular spacing rule (see the detail) the top of the block should match the line at the "2."

Part two of this sketch shows a *story pole* being used to check for proper course height. To fashion one, select a straight 2×4, and scribe a mark every 8 inches. The top of each block should align with one of the marks.

No matter which tool you use to check course height, if the block is too high, tap it down into the mortar with the trowel handle. If too low, pick it up, place more mortar on the bed, and re-lay the block. *(continued)*

Building Block Walls *(continued)*

6 Once you've laid the first block, place a level along its length to make sure it is exactly level. If it isn't, press down or tap the high end until the block is level.

7 Now check the block for plumb (true vertical) by laying the level across the block, as shown. If the block isn't plumb, press with the heel of your hand or tap with the handle of your trowel to adjust it.

8 Before you lay the second block, apply a mound of mortar to its flanges (or ears) while it is standing on end. This mortar will form the vertical joint between the first two blocks in the wall. Now lower the block into place and repeat steps 5 through 7.

9 After you have laid the first course (here we show a corner situation), check your work for square, level, and proper alignment. (A framing square and a 4-foot level are the correct tools for these tasks.) Make any necessary adjustments by gently tapping or pushing on the blocks. If the mortar in the joints breaks up when you adjust a block, take up the block, remove the mortar, and repeat the process.

10 Apply a 1-inch-deep layer of mortar along the edges of the first block in the second course. It's not necessary to put mortar on the *webs* on the inside of the block.

As you place the second corner block, align its outside corner with the corner below it. Press the block just enough so its weight compresses the mortar to a ⅜-inch joint.

Continue laying blocks until you have a *corner lead* as shown in the detail. Then check your progress by laying your level or a straightedge diagonally alongside the end blocks as shown. If the spacing is as consistent as it should be, all the blocks' outside bottom corners will align. When you've reached a height of four or five courses in one corner, build a second corner along one leg of the first corner.

11 If your project calls for tying the new wall to an existing one, this sketch shows you what to do. Starting with the second course and repeating with every other course, knock a fair-sized hole through the existing structure's corner block, stuff old newspapers into the cavity of the block below the one with the hole in it, and place a length of ⅜-inch re-rod (bent into an S shape) into the hole. Now trowel mortar into the cavity containing the re-rod. Then, stuff newspapers in the cavity under the exposed end of the re-rod, lay the second course of block as explained earlier, and partially fill the cavity with mortar.

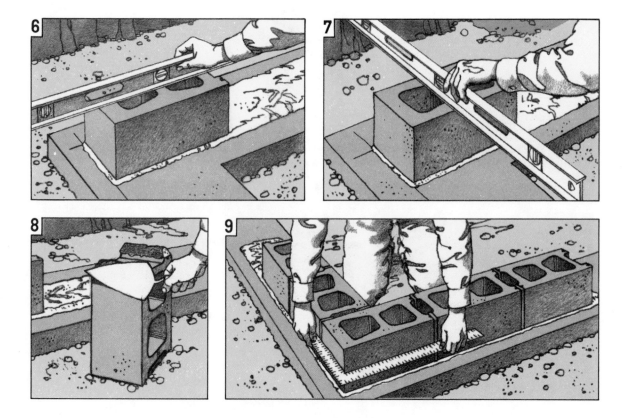

WEBS

CORNER LEAD

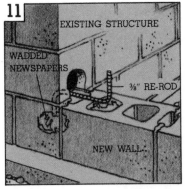

EXISTING STRUCTURE

WADDED
NEWSPAPERS

⅜" RE-ROD

NEW WALL

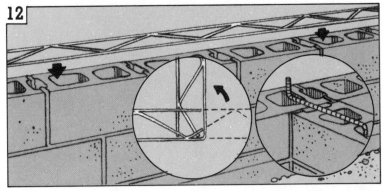

12 In situations such as foundation walls and retaining walls, where the wall must withstand considerable lateral pressure, it's best to beef up the wall with truss-type reinforcing wire, embedded in the mortar of every other horizontal joint. Overlay the ends of the truss sections by at least 6 inches. The first detail shows how to position the reinforcing wire when turning a corner. And the second detail depicts how to tie intersecting walls together with ⅜-inch re-rod.

LINE
BLOCKS

13 Once you have two corner leads in place, you're in a position to fill in between them with *stretcher blocks*—those with ears on both ends. To do this, first stretch a mason's line between a pair of *line blocks*, which hold the line in place. Hook the line blocks around the corners as shown here. The mason's line should align with the top of the blocks in the course being worked on and be about ⅟₁₆ inch away from the blocks' outer edges.

With the mason's line in place, begin setting the blocks as explained earlier. Check the line often to make sure no blocks or pieces of mortar are touching it to push it out of alignment.

JOINT
STRIKE

14 Butter both ends of the final or *closure block* and lay it to complete the course. If some of the mortar falls off the flanges, lay the block anyway. Fill the joints by tucking mortar in place from the sides with your trowel.
(continued)

Building Block Walls *(continued)*

15 In some situations you may have to cut concrete blocks to fit. To do this, first place the block on sand or loose soil and, using a hammer and brick set, mark where you want the cut to be on both sides of the block. Then work along the line again, hammering somewhat harder and moving the brick set each time you rap it. Continue until the block breaks along the line. For precision cuts, use a circular handsaw with a diamond or abrasive blade made for cutting masonry. Cut only dry block this way.

16 Plan to place openings for doors and windows so you won't have to cut blocks to fit around them. For doors, install the sill and jambs or doorframe before you begin laying block. You can use metal units designed for use in masonry walls or build wooden frames. Put in window frames when the wall reaches the appropriate height. In this sketch, the blocklayers are setting a precast lintel in place to provide adequate support over the doorframe.

17 How you cap off a block wall depends on its function. If your wall will serve as a foundation for a building, you need to embed anchor bolts in mortar in the cores of the top course of blocks. This ties down the *sill plate,* the wood member on which the house framing rests. So as you lay the next-to-last course, provide a wire mesh bed beneath all of the cores that you intend to fill with mortar.

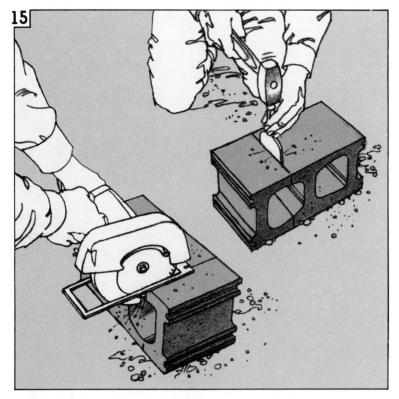

If the foundation wall will later support brick or stone veneer, lay a course of corbel block at the appropriate height, and repeat the above procedure. Cap exterior walls, such as garden walls, with solid 2-inch blocks.

18 Finish each joint after you have laid two courses of block above it. This provides enough time for the blocks to absorb some of the moisture in the mortar, making it firm enough to work the joint without harming it. With above-grade blocks, tool the *head* (vertical) joints first and then the *bed* (horizontal) joints with a joint strike. For blocks that will be below grade, simply strike off excess mortar with your trowel.

19 Soon after finishing the joints, use a soft-bristled brush to whisk away bits of mortar left at the edges of the joints.

20 Sealing a block wall to moisture involves a couple of two-coat applications. First, apply a ¼-inch coat of mortar to blocks that will be below grade. Scratch this layer, while wet, with a wire brush, allow the mortar to dry for 24 hours, and then apply a second ¼-inch coat and allow it to dry. This two-step process is called *backplastering*.

21 After letting the backplastered walls cure for several days, apply two coats of bituminous waterproofing with either a dash brush or a roller. This material will harden into a tough, tar-like coating.

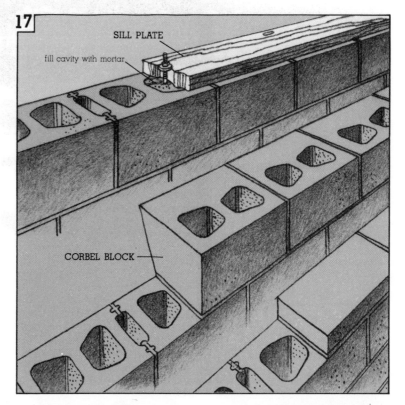

17 SILL PLATE

fill cavity with mortar

CORBEL BLOCK

18 HEAD JOINT

BED JOINT

19

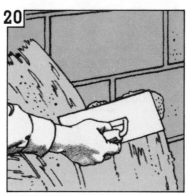

20

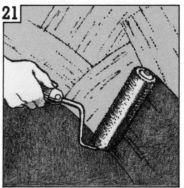

21

Erecting Brick Walls

No matter what brick project you're contemplating—privacy screens and fences, fireplace facings, brick veneers, and all the rest—they're all basically walls in various configurations. That's why on the following few pages we take you step by step through the wall-raising process for both single-tier and two-tier walls. Techniques used with brick walls are similar to those used with concrete block walls (see pages 54–59), but differ in several ways, too, as we show here.

As with any other project, planning is important when building with bricks. Here are some things that need

doing before you begin the actual construction.

First: Realize that local building restrictions often apply to some masonry projects—most particularly to brick walls—and prescribe minimum dimensions and other specifications. So be sure to check with municipal officials early on to verify that your plan is acceptable.

Second: Like most masonry items, bricks are heavy—about 5 pounds each. So, to resist settling and cracking, brick walls (and other brick projects) must rest on concrete footings that reach below the frost line. In general, the footing should be twice as wide as the wall it supports, and its vertical depth should be equal to the thickness of the finished wall. (See pages 18–19 and 24–25 for details on building footing forms.)

However, in at least one case—brick veneer on a house—the bricks rest on concrete blocks or on a metal angle or concrete grade beam that's attached to a concrete wall. See the sketch at right.

Third: Using the sketch below, select the *bond* (or brick pattern) and the mortar joint configuration you find most attractive. (Note, too, the names of bricks when laid in certain positions.)

Fourth: Review the information on how to figure material needs (page 43) and how to mix and work with mortar (pages 50–51).

Single-Tier Brick Veneer

1 Begin by excavating a trench a foot or so deep and at least 18 inches wide along the foundation as shown. If you'll be bricking

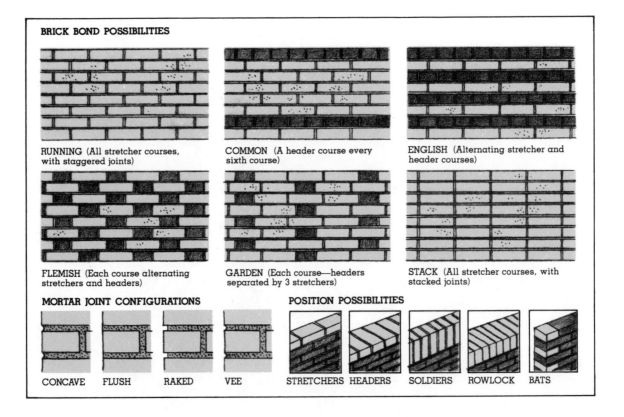

BRICK BOND POSSIBILITIES

RUNNING (All stretcher courses, with staggered joints)

COMMON (A header course every sixth course)

ENGLISH (Alternating stretcher and header courses)

FLEMISH (Each course alternating stretchers and headers)

GARDEN (Each course—headers separated by 3 stretchers)

STACK (All stretcher courses, with stacked joints)

MORTAR JOINT CONFIGURATIONS

CONCAVE FLUSH RAKED VEE

POSITION POSSIBILITIES

STRETCHERS HEADERS SOLDIERS ROWLOCK BATS

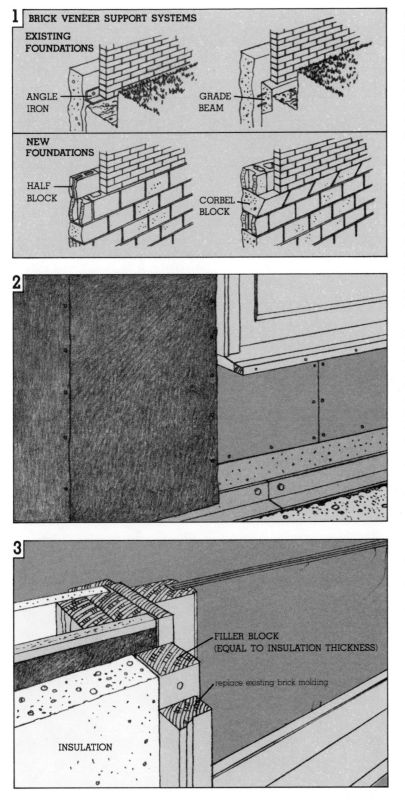

1 BRICK VENEER SUPPORT SYSTEMS

EXISTING FOUNDATIONS

ANGLE IRON

GRADE BEAM

NEW FOUNDATIONS

HALF BLOCK

CORBEL BLOCK

2

3

FILLER BLOCK (EQUAL TO INSULATION THICKNESS)

replace existing brick molding

INSULATION

around any windows or doors, use a folding rule to determine at what level the first course should be so that a rowlock course can be positioned conveniently beneath the opening (see page 64, sketch 13). Then snap a level chalk line 4 to 6 inches below grade to establish where the top of the angle iron or grade beam will be.

To fasten an angle iron to either concrete block or a poured concrete foundation wall—a 4×4× ⅜-inch iron works well—first drill holes every 2 feet in the wall and insert metal wall anchors sized to accommodate ⅜×4-inch lag screws. (With concrete block walls, drill into the mortar joints at the appropriate height.) Then with a helper, lift the angle iron into place and drive the screws into the anchors.

To tie a grade beam to an existing foundation wall, drill holes and insert short lengths of re-rod before placing the concrete grade beam. And when you place the concrete, use at least two lengths of ½-inch re-rod, positioned horizontally, to add strength to the beam, which should be at least 6 inches wide and 12 inches deep.

Obviously, with new construction, you would use the top of the foundation to support the veneer or a course of corbel block a few inches below grade (also shown).

2 Remove the siding on the wall section(s) you plan to veneer and cover the wall(s) with 15-pound builder's felt so it overlaps the edge of the steel angle or grade beam. Nail a redwood extension to windowsills so they reach at least ⅜ inch beyond the new veneer to facilitate drainage.

3 If you live in a cold climate, this is a good time to add 1 to 2 inches of rigid insulation between the existing wall and the new veneer. If you decide to insulate, remove the openings' molding, trim out the doors and windows with redwood filler blocks of the appropriate thickness, as shown, and reattach the molding. *(continued)*

Single-Tier Brick Veneer

(continued)

4 Before mixing the mortar, dry-lay the first course of bricks, and make any necessary position adjustments to avoid having to cut any more bricks than necessary. At about the same time you're mixing the mortar for your job, soak or sprinkle several bricks liberally with water to prevent them from absorbing moisture too rapidly from the mortar. Then, on one end of the angle iron, throw an inch-deep bed of mortar long enough to lay two or three bricks. With the tip of your trowel, furrow the bed as shown. (Furrowing spreads the mortar to help improve the bond between the bricks and mortar.)

5 Gently press the first brick into place, keeping in mind that you want a ½-inch space between the brick and the wall surface, as well

as a ⅜-inch joint beneath it. The ½-inch air gap helps insulate the wall and allows any condensation that may form in the space to drain. Check for proper spacing with a folding rule. (A better way to check spacing is to construct and secure 1×4 story poles—complete with the appropriate markings and replicating those on the spacing rule—to the building at the ends of the work.) Trim away excess mortar forced from the new joint. Return the excess to your mortarboard or place it farther down along the angle iron.

6 After you have laid three or four bricks, first measure the thickness of both the *head* and *bed* joints. They should all be ⅜ inch thick. If they aren't thick enough, take up the bricks and put in more mortar before re-laying them. Now check the bricks' placement with a level to make sure the tops are even with each other and level. Then check the face of each brick to

make sure it is plumb. And finally, lay your level horizontally along the face of all the bricks to check for proper alignment. Adjust any offenders, even if this means re-laying them.

(NOTE: Be sure to do all this checking immediately after laying the first few bricks so you can adjust their positions as needed. If you wait too long, the bricks will have absorbed too much of the moisture in the mortar to allow you to move the bricks without the mortar crumbling. If this happens, remove both the bricks and mortar, and start over.)

7 Continue laying bricks in the manner described in steps 4 through 6 until you have built a 5- or 6-course-tall lead at each end of the angle iron or grade beam. Check often to make sure all units are level, plumb, and aligned. Note the flashing between two of the courses in the leads (see also the detail). You should place the flashing one or two courses above

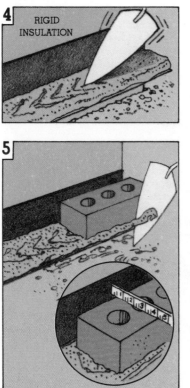

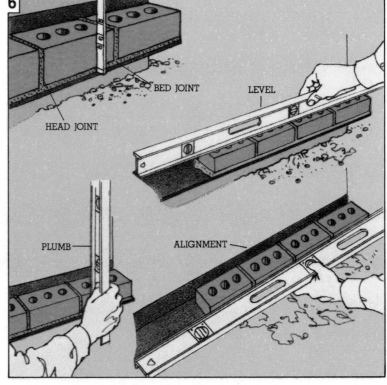

FLASHING

CORNER LEAD

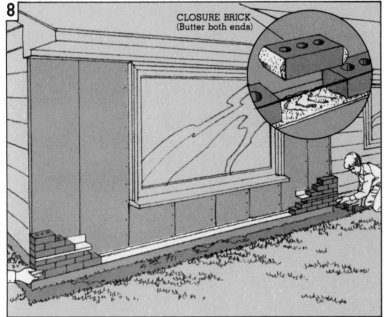

CLOSURE BRICK
(Butter both ends)

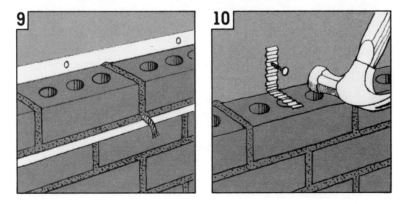

the final grade to help channel water away from the foundation.

8 Now stretch a mason's line between the corner leads as shown. Position the line blocks so the mason's line aligns with and is about 1/16 inch out away from the top of the first course of bricks. Lay the remainder of the bricks in the course. Then move the line blocks up one course and continue laying to the line.

9 When you reach the course that will contain the flashing, nail the flashing in place, overlapping the sections you placed while building the leads, then proceed to lay the bricks.

As you lay the bricks in this course, you also need to provide for drainage of any moisture that builds up behind the veneer. To do this, insert short pieces of 1/4-inch sash cord—at 2-foot intervals—at the bottom of header joints (see sketch). Make sure the cords extend all the way through the veneer. As soon as the mortar stiffens enough to hold its shape, pull out the cords to make *weep holes.*

10 To tie the veneer to the existing wall, nail brick ties every 32 inches horizontally and 16 inches vertically. Stagger the rows across the wall so the ties are no more than 24 inches apart. For strength, use 8- or 10-penny ring-shank nails. As you brick over the ties, embed them completely in mortar.

(continued)

Single-Tier Brick Veneer *(cont.)*

11 Using the tool needed to produce the mortar joint you choose (the striking iron shown here makes the familiar concave joint), strike the head (vertical) joints first, then the bed (horizontal) joints. **NOTE:** Don't allow the mortar to stiffen too much before striking the joints or you may not be able to work it.

12 After striking the joints, let the mortar set up for a while, then remove the burrs and crumbs of mortar squeezed out at the sides of the joints. A stove brush does this job nicely, but you also can use the edge of a mason's trowel.

13 To brick up the bottom of a windowsill or to cap off a veneer that extends only partway up a wall, set *rowlocks* as shown here. Cut the bricks to the desired shape either with a brick set and hammer or with a mason's hammer. As you lay the bricks, you may have to adjust the joints to make the rowlock fit the opening.

14 To provide support for bricks that will reside over doorways and windows, secure a 4×4×¼-inch angle iron as shown. Its ends should rest on a course that is flush with the top of the sides of the opening and overlap the sides by at least 4 inches for plenty of support.

15 If you brick all the way to the soffit, cover the gap between the soffit and the top course of brick with molding. This serves the two-fold purpose of giving the job a finished look and keeping out moisture and insects. *(continued)*

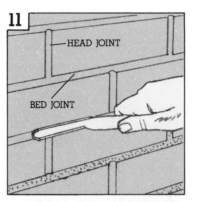

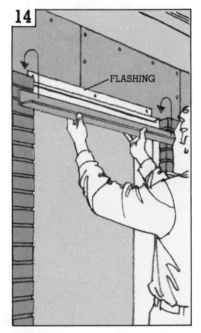

16 Use butyl or latex caulking to seal all seams where brick molding and masonry meet. Fill gaps wider than ½ inch with insulation or oakum before caulking the joint.

Two-Tier Brick Garden Walls

Many of the steps needed to construct brick garden walls closely parallel those described for single-tier brick veneer walls. So before beginning, please read steps 4-8 on page 62 and steps 11-12 on page 64. To learn about footing requirements, study the information on page 24. Also review the planning pointers on page 60.

1 After digging a trench the length of the wall, as deep as the frost line in your area, and twice the width of the wall, place concrete to within a few inches of the top of the excavation. Allow the concrete to set up for several days.

Then, with a helper, snap chalk lines along the footing to define the perimeters of the wall, which should be centered on the footing. Now, begin building the wall as discussed in steps 4-8 on pages 62-63 and steps 11-12 on page 64.

As you're laying up the bricks, embed corrugated metal brick ties in the mortar about every 12 inches in every third or fourth course.

2 Once the wall reaches the desired height, cap it off with *coping*. Common coping materials include solid header bricks and bricks placed rowlock fashion, concrete blocks, formed concrete, and limestone. Slope the coping to allow for adequate drainage.

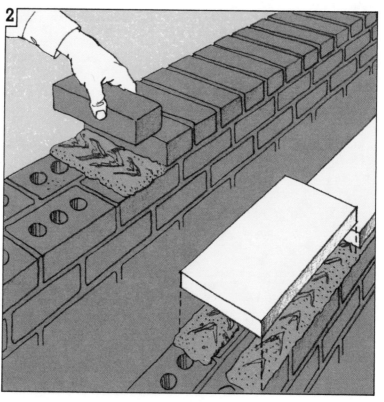

Installing Lightweight Veneers

What's so special about today's generation of lightweight brick and stone veneers? Well, to start with, many of them look as attractive as their more bulky counterparts.

Also, their minimal weight opens up new do-it-yourself possibilities unheard of only a few years ago. Imagine being able to face a fireplace with stone or a chimney with brick without having to worry about building footings, adding additional bracing to support the extra weight, or installing lintels over openings. And if these things weren't enough, the material itself costs less than the conventional brick and stone.

On this and the next page we show how to lay up both lightweight brick and stone. Note, however, that several manufacturers make these materials. So before beginning your installation, read the directions on the package

of the product you buy. (To estimate your material needs, see page 43.)

Lightweight Brick

1 Start by measuring in inches the width of the wall to be covered. Then divide by the length of each brick plus one joint to determine the number of full bricks you'll need. If the result is anything but a whole number, you'll need to start with part of a brick. The length of the first and last bricks in the first course should be the same, so divide any fraction of a brick by two. If your project calls for going around one or more corners, you may have to score and cut corner pieces to fit.

Now spread a ⅛-inch layer of adhesive over a 2×4-foot area, starting in an upper corner of the wall. (To prepare outside walls, first nail a moisture barrier of 15-pound roofer's felt and a layer of expanded metal lath to the walls with roofing nails. Then trowel on a coating of mortar, working in 2×3-foot sections.)

2 Press each brick into the adhesive (twisting it slightly to ensure a good bond) and align it with a taut guide string as shown. Leave ⅜-inch spacing between units for normal-looking joints. To cut bricks

to fit, use a hacksaw frame fitted with a masonry cutting rod.

3 Fill the joints between bricks with more of the mortar adhesive. Use either a pencil-type brush or grout bag with the appropriate-sized tip. Avoid smearing the faces of the bricks with mortar, and wipe up spills immediately. Once the joints have dried, coat the veneer with a sealer.

Lightweight Stone

Lightweight stone veneers may be made of natural stone or cast from concrete and processed to have the natural-looking colors and textures. These veneers range from ½ inch to 2 inches in thickness. With some of these products, you can use an adhesive to secure pieces of the material directly on a finished interior wall, as was just described for brick veneers. With others, however, the manufacturers recommend going with expanded metal lath and mortar, as shown here. (See page 43 for help in determining your materials needs.)

1 For interior walls, start by nailing expanded metal lath to the wall. With exterior walls, nail up roofer's felt first and then metal lath. Now trowel a ¼-inch coat of mortar over the lath (see pages 50–51 for how to prepare mortar).

After the mortar has set up slightly, roughen the surface with a rake-like tool, scratching to a depth of ⅛ inch. Let this coat dry and cure for 48 hours before applying stone. Meanwhile, lay the stones out on the floor as they will be positioned on the wall.

2 Cover the scratch coat with a ½-inch layer of mortar, working with an 8-square-foot area at a time. Also apply a thin layer of mortar to the back of each piece of stone veneer just before positioning it. Press the stone into the mortar bed, moving it back and forth and rocking it slightly to form a thorough bond. Place corner stones first, then work toward the center of the area you are veneering. (See pages 46–47 for how to cut or break stones to fit.)

3 As soon as the mortar has set up sufficiently, strike the joints. Then brush away the crumbs and burrs at the joints' edges.

SCRATCH COAT

METAL LATH

MORTAR

SCRATCH COAT

STONES READY FOR PLACEMENT

Applying Stucco

Long known for its durability and weather resistance, stucco is an excellent alternative to brick or stone for exterior walls and for certain indoor applications as well. And even though its popularity has dropped somewhat in recent years, you can find all kinds of testimonials—mostly older homes—to its long-lasting good looks.

Though most people recognize stucco when they see it, not many know what it really is. A stucco finish is nothing more than two or three thin coats of mortar that is one part masonry cement to three parts sand, with just enough water added to achieve a putty-like consistency. The techniques for applying it haven't changed much over the years. Now, however, expanded metal lath rather than wood strips and wire often help tie the stucco to its backing, which generally is wood or concrete. If left un-

tinted, stucco will dry to a medium-gray color, which many people find quite attractive. But if you'd rather, it's possible to color it, too. You can add an oxide pigment to the finish coat of mortar, or stain or paint the surface after the topcoat has cured.

If you mix in pigment, be certain that you carefully measure the amount of coloring agent, and keep track of the proportions of all of the ingredients so that succeeding batches are consistent in color. If you paint or stain stucco, select a material that is designed for covering concrete or masonry products. Your local paint dealer can help in this selection.

You can even make brilliant white stucco by mixing white portland cement, lime, and white silica sand for the finish coat. It is beautiful, but expensive.

1 The wide range of finishes possible with stucco makes it one of the most versatile wall coverings. To achieve a *smooth, plasterlike appearance,* you need to trowel the final coat several times as it becomes progressively stiffer. For a *swirled texture,* trowel the mortar just once, using an arc motion, and allow the resulting pattern to remain.

For a *wavy, striated surface,* draw a brush lightly over a smooth-troweled surface after it has hardened slightly. A soft brush will produce a fine wavy pattern; stiffer brushes, predictably, will give it a coarser look.

To *stipple* the top coat, hold a whisk broom at an angle to the wall and pat the surface with the ends of the bristles to make an irregular pattern. For a *travertine finish,* spatter on a coat of thin mortar of a contrasting color and trowel slightly after it has stiffened. And to *imprint,* press leaves or other patterns into the soft mortar and trowel the surface lightly.

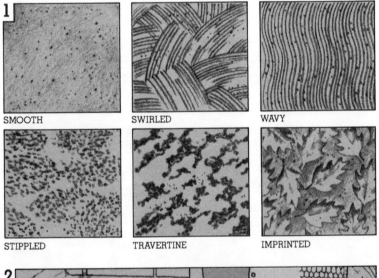

SMOOTH SWIRLED WAVY

STIPPLED TRAVERTINE IMPRINTED

2 How you prepare a wall for stucco depends on the material the wall is made of. With concrete walls, preparation involves nothing more than wetting the surface to be stuccoed shortly beforehand. To apply stucco over a wood-framed wall, however, first nail up 15-pound roofer's felt, then cover that with expanded metal lath. Now trowel on the first ¼- to ½-inch layer of mortar (the "scratch coat"). If you've nailed up lath, force the scratch coat into the mesh so some of it hangs through slightly. Here, the idea is to cover the mesh.

3 Allow the scratch coat to harden slightly, then roughen it with a plasterer's rake or a homemade tool such as the wood-and-nail scratcher shown in the sketch. Scratch the entire mortar surface to a depth of about ⅛ inch, running the scratcher in long lines along the surface.

4 As with all concrete, slow, damp curing provides the most strength. Allow the scratch coat to cure for 24 hours, keeping it damp by misting with a garden hose periodically during that time.

5 With a steel trowel, lay onto the dampened scratch coat a ⅛- to ¼-inch-thick finish coat. Finish to the texture of your choice and allow the stucco to cure for several days, again misting the surface occasionally to aid in the curing process. This step is critical, especially when working under a hot sun or in a strong wind. Complete the project by painting the stucco, if desired, and by caulking around doors and windows to prevent moisture from entering.

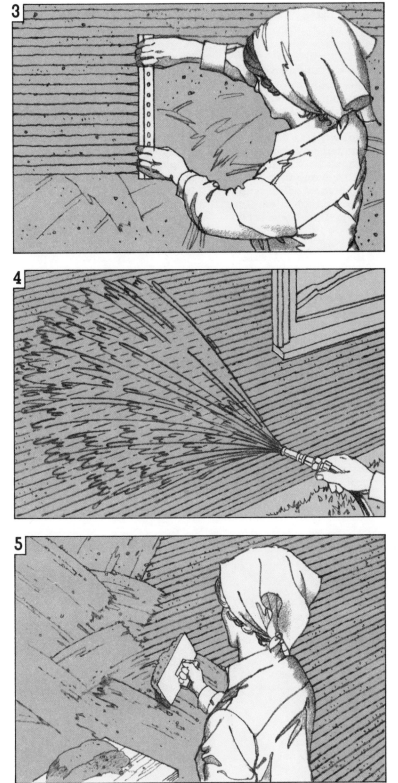

Solving Masonry Problems

Although masonry projects promise permanence, even the best-built ones require occasional repairs. That's why on this and the next three pages we show you how to deal with the things that can go wrong—things such as cracks, broken blocks, leaking, even efflorescence. We also show you how to make openings in masonry, a problem unless you know how to do it.

Cracking, the most common masonry malady, results from uneven settling of footings or from expansion and contraction triggered by temperature changes.

You can diagnose the cause of cracking by careful inspection. *Expansion cracks* usually occur with uniform width and often follow joints between masonry units. *Settling cracks,* however, taper along a vertical path, being widest at the top and ending as hairline cracks near the wall's bottom.

Horizontal cracking, another problem, may appear in basement walls made of con-

crete blocks. Generally, pressure from backfill soil pushing in from the outside is the culprit here.

The steps for repairing cracks in brick structures and block walls (see sketches 1–6) are identical, as are those for replacing cracked bricks and blocks (see sketches 1 and 2 on pages 71–72). With horizontal cracks, the backfill must be dug out and the affected blocks removed and re-laid. This major overhaul job normally requires a professional contractor.

Inadequate drainage around footings and behind basement walls and floors causes another masonry-related problem: wet basements. Waterproofing outside foundation walls will help. But relieving severe seepage may require hiring professionals who are well equipped to make major modifications.

For small repairs, a variety of easy-to-use materials are available, including mortar, special latex- and epoxy-based patching materials, and caulking compounds.

NOTE: When working on wall sections more than 5 feet above the ground, put up a scaffold or make a platform by laying planks across sawhorses.

Repairing Cracks

To repair cracks caused by settling, it's imperative that you wait until the movement stops before beginning. This may be as long as a year after the first sign of cracking appears. To determine whether settling is still occurring, bridge the crack with a piece of tape and check it occasionally for twists, a tear, or looseness.

1 With a heavy hammer and a cape chisel, remove mortar from the cracked joint to a depth of ¾ to 1 inch. Because chips will fly as you work, wear safety goggles and heavy gloves. If the joint crumbles easily all the way through the wall, you should tear down the whole section and re-lay the bricks with new mortar. Tuck pointing by itself will not add strength to weak masonry joints.

2 Using the point of the chisel or the tip of a mason's trowel, scrape

away patches of mortar that remain after chiseling. Then briskly sweep away sand or dust with a whisk broom.

3 Wash away any remaining debris with a strong jet of water. Wetting the surface also improves the bond between the masonry units and the mortar you use for repairing the affected joint. If the wall has been dry for a long time, let the moisture soak in, rewetting the surface every 15 minutes and just before you apply the mortar.

4 With a caulking trowel, force fresh mortar into the cleaned

joints. For this use, prepare a mix of one part portland cement to two parts masonry sand and enough water to form a putty-like consistency (see pages 50–51 for details on mixing and testing mortar). Use a hawk to carry and hold the mortar as you work. Fill the head (vertical) joints first and then the bed (horizontal) joints.

It probably will take some practice to tuck point the joints without dropping mortar or smearing some of it on the masonry units. Use a damp sponge to wipe away unwanted mortar while it is still wet. Brush the joints to remove mortar crumbs.

5 Now strike the repaired joints so they match the shape and the depth of the other joints in the wall. Strike the head joints first, then the bed joints.

6 After striking the joints, let the mortar set up somewhat. Then, with a stove brush, remove any mortar crumbs that squeezed out from the edges of the joints during the striking operation. And within 24 hours, scrub mortar stains from the units.

Replacing Bricks and Blocks

1 Just one broken unit can practically ruin the appearance of a brick or concrete block wall. The best remedy is to remove the broken unit and replace it. To do this, carefully chip away the mortar and remaining parts of the damaged piece of masonry. With bricks, take out the entire unit. With concrete blocks, unless the block is completely fractured, remove only the damaged face and leave the rest. *(continued)*

Replacing Bricks and Blocks *(continued)*

2 After brushing away the debris and wetting all surfaces of the cavity, lay a bed of mortar ½ to 1 inch deep on the bottom of it. Butter the top and ends of the brick or 2×8×16 block you'll use to replace the broken one. Slide the piece in as carefully as you can and adjust its position so joints surrounding it match adjacent ones. Force additional mortar into the joints as needed to fill them completely. Finish the joints and clean up the repaired area as described on page 71.

Stopping Leaks and Waterproofing

1 Stopping leaks in concrete or block walls or floors calls for working under damp or even wet conditions. But with hydraulic cement you can make a repair. These products are usually dry mixes you add water to, then apply; or they can be putty-like formulations you can use as is. Shape the material into a plug, then squeeze it into the opening till full. As it dries and cures, the plug will expand and tighten.

2 Drying up wet or leaky basement walls requires that you stop moisture on the outside. Excavate a 2- or 3-foot-wide trench along the foundation and all the way down to the footings.

Scrape all soil off the wall surface and wash it clean. When dry, apply two coats of tarlike bituminous sealer, or backplaster with two coats of mortar (see page 59). Then coat with bituminous sealer.

Before backfilling the trench, install sloped drain tile and ¾-inch gravel along the footings (see page 12) to create a drain line that empties at a lower elevation.

Removing Efflorescence

1 Where any crystallized surface stains (efflorescence) are light, you can remove them with water and a brush. If that doesn't work, wet down the wall and scrub it with a solution of 1 part muriatic acid to 15 parts clean water in a plastic bucket. Always add acid to the water when mixing. And wear rubber gloves and other protective clothing while using the acid solution. Flush the wall with water when finished. Repeated acid cleanings may be necessary until the efflorescing stops.

Making Openings in Masonry Walls

1 To make any opening in a masonry wall, indicate the opening's perimeter by marking it with a pencil or by snapping a chalk line. Use a level to make sure the sides are plumb before scoring along the lines. For small cutouts, such as an electrical outlet, drill several holes around the perimeter, then pound on the section inside the drilled holes. Eventually the unit will yield.

With large openings, such as those for doors or windows, use a portable circular handsaw with a masonry blade to score the surface to a depth of ½ inch. Then chip out one corner first with a hammer and chisel until the first two or three whole units are gone. Now you can use a heavier hammer or sledge to continue breaking away pieces of block as shown.

Work carefully to avoid damaging blocks outside of the scored lines. Also remove units at the top of the opening and install either an appropriate-sized steel angle or a concrete lintel. Re-lay the blocks above it with new mortar (see pages 50–51 and 54–59 for details on mixing mortar and laying blocks).

Making any opening of this type involves toiling hard under dirty, dusty conditions. So wear gloves and eye cover as you work.

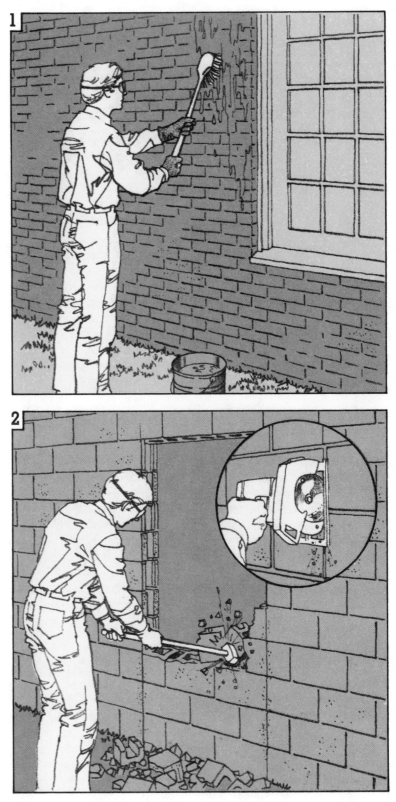

SPECIAL-EFFECT PROJECTS

In the first two sections of this book, we presented the nuts-and-bolts information on how to work with concrete and with brick, block, and stone. But here the emphasis changes because we know you're probably eager to put your newfound knowledge to work.

In this final section, then, you'll find ten nifty projects that are guaranteed to draw rave notices from all who see them. You'll find out how to erect a splendidly simple lawn bench, build a concrete block privacy screen, face a fireplace with lightweight stone veneer, construct a handsome brick chimney, and more. Step-by-step drawings and instructions accompany each project, as do cross-references to appropriate pages in the first two sections.

But don't confine your thinking to just the projects on the next 18 pages. If you can't find one that suits you, sit down and rough out a creation of your own. Or drive around town looking for good ideas. Or leaf through various shelter magazines or books. The information you've gained will enable you to successfully tackle any project you run across.

Obviously, some projects require a much greater investment of time and energy to complete than others. So if your plans are grandiose, but you're a little short on time, you may want to consider dividing the project into a number of manageable segments done over several weeks. For example, to put in the brick fireplace on pages 88–89, you could excavate and pour the footing one weekend, build the chase another time, and finish up by doing the actual brickwork still another time.

Brick and Redwood Bench

This simply styled outdoor "rest awhile" combines two classic materials in a way that's sure to be a winning combination in any setting. And as you can see, you needn't be a master mason to build it, either.

As you plan for this project, give plenty of thought to its location. Perhaps you will use it to expand a patio, dress up an entry, or provide a special place just to relax and enjoy those flower beds.

You'll also need to dimension the project to suit your situation. Typically, seating height measures from 16 to 20 inches, and the width from 16 to 24 inches. As for length, a bench of this type will easily span 8 feet.

1 Stake out the perimeters of the supports, excavate to below the frost line, and pour concrete footings to within a few inches of grade. Lay brick support walls to a height that's three or four courses higher than the seat level (see pages 42–43, 50–51, and 60–65 for brick-laying specifics). Allow the mortar to cure for a few days, then with a masonry bit, drill holes for the masonry anchors. Attach the steel angles with ¼-inch lag bolts.

2 To assemble the seat section, cut redwood 2×4s to the same length and nail them and redwood spacers together with 8-penny galvanized nails. To prevent boards and spacers from splitting, blunt the nails before driving them. Place a set of spacers at each end of the seat and midway between.

3 Bore pilot holes in the seat and fasten it to the steel angle brackets with ¼×2-inch lag screws.

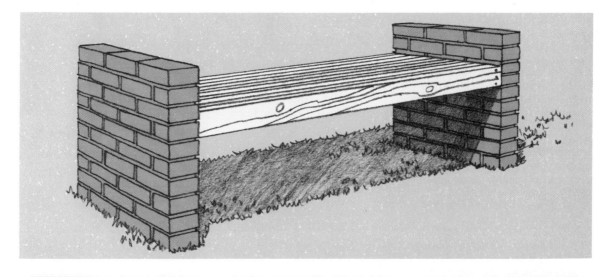

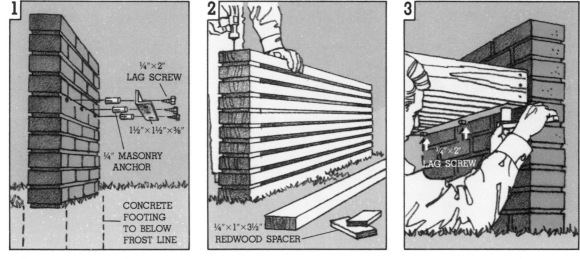

1 ¼"×2" LAG SCREW — 1½"×1½"×⅜" — ¼" MASONRY ANCHOR — CONCRETE FOOTING TO BELOW FROST LINE

2 ¼"×1"×3½" REDWOOD SPACER

3 ¼"×2" LAG SCREW

Fashionable Mailbox Standard

Here's an exquisite alternative to the usual, uninteresting way of putting up a mailbox. This standard makes a quiet statement about how positive the people who erected it feel about their home—how it speaks of strength and permanence.

Before starting construction, consider the standard's placement carefully. You'll want it far enough from your driveway's side to minimize the possibility of accidents, but not so far that fetching the mail becomes a chore. Remember, it has to be within the mailman's reach, too. And keep in mind that the mailbox itself must be "U.S. Post Office-approved" and must stand at regulation height. Generally, the bottom of the box should be from 42 to 46 inches above street level. But to be sure, check with local postal officials.

NOTE: If there's brick on your home, consider using the identical style and size for this project to tie the two together visually.

1 Stake out the perimeters of the project, excavate to below the frost line, and pour a concrete footing to within an inch or so of grade. After reviewing the material on pages 43, 50–51, and 60–65, lay the first course of bricks for the standard. (Starting with the first course and repeating every four courses thereafter, use metal brick ties as shown to bond the single-tier section to the two tiers in the rear.) Continue laying additional courses until the pedestal is nine courses above grade.

2 Divide the tenth course above ground level into two sections. Continue laying the rear part for another ten courses. Build the front section as a quarter-circle on a radius of about 12 inches. The radius rotates along the upper front corner of the ninth course of bricks above ground level (see dotted lines in the sketch). You may wish to cut a stick to the radial length as a guide for marking and cutting bricks to fit (see page 47 for information on how to cut bricks). Complete this section of the pedestal by carefully laying a course of rowlocks along the curve, taking care to maintain uniform joint width. Use a solid brick to cap off the arch. If your local brick supplier doesn't carry solids that match your bricks' style, go with what you have, filling any holes neatly with mortar.

3 Cut pieces of 1×4 and nail them together to provide a form that fits snugly around the top of the column and extends 2 inches above it. After making sure the form is level, fill it with mortar. Shape the top of the cap as shown and trowel smooth several times for a good, weathertight finish.

4 Let the mortar set and cure for at least two days before drilling holes for masonry anchors in the rear column and at the top of the curved section. Attach a wooden ledger to the rear column, pound a masonry anchor into the hole in the front section, and then nail a support shelf to the ledger. Secure the shelf at the front with a lag screw driven into the masonry anchor. Now secure the mailbox to the support shelf.

Complete the project by adding your address number to the side of the standard.

Serpentine Planter Wall

This lovely, meandering brick wall transforms an ordinary planter area into a virtual work of art. The brick provides an interesting change of pace in both texture and color that will complement the overall appearance of your plants and your home.

You'll be surprised by how easy it is to build this gracefully curving wall along your favorite garden spot. The biggest challenge is making the template, a task we explain fully in sketch and caption 1.

1 From a 4×8-foot sheet of ½-inch plywood, saw out 24×48-inch and 24×96-inch pieces and nail them together end to end with cleats as shown. Lay the plywood panel on a flat surface, then establish two base lines at right angles. Draw one 2 inches in from one edge of the template. Draw the second along the end of the panel, using a carpenter's square to make sure the lines are perpendicular to each other.

Measuring from the base lines, locate the radius centers shown and mark their locations with a pencil or grease marker. With a string and pencil, draw the radii on the template. Make sure you don't knock the template out of position as you work. After completing the pattern, cut out the area between the lines.

2 Stake the template in place on the ground, making sure you maintain an 8-inch opening the entire length. The easiest way to ensure this is to maintain 24 inches between the outside edges of the template. Now with a spade dig a

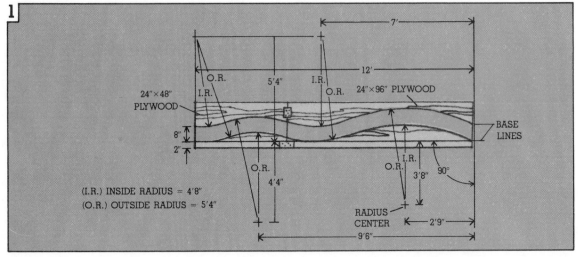

1

24″×48″ PLYWOOD

24″×96″ PLYWOOD

O.R.
I.R.
5'4"
I.R.
O.R.
7'
12'

8"
2"

BASE LINES

O.R.
4'4"
I.R.
O.R.
3'8"
90°

RADIUS CENTER
2'9"

(I.R.) INSIDE RADIUS = 4'8"
(O.R.) OUTSIDE RADIUS = 5'4"

9'6"

12- to 18-inch-deep trench along the curve of the template. Fill this trench with concrete to within a couple of inches of the grade. Allow it to cure for a few days.

3 Use the template as a guide for laying the bricks, too. First make a 1-course dry run to see how well full bricks fit and to adjust the width of the vertical joints between them. Stagger the two tiers of bricks as shown.

After reviewing the bricklaying information on pages 42–43, 50–51, and 60–65, lay the first course of bricks. Place metal ties across the tiers at 24-inch intervals to tie the two tiers together. Do this every couple of courses.

As you lay the course just above final grade, make weep holes through the wall (see page 63 for how to accomplish this). Put these 12 or 18 inches apart. Check the wall for plumb often so it will conform to the shape of the template as you lay additional courses.

4 Cap the wall with a rowlock. At the corners, position the bricks as shown. For a more finished-looking corner, miter-cut the bricks and place them as shown in the detail. (See pages 46–47 for details on cutting bricks.)

Allow the mortar to cure for a few days. Then coat the foundation wall with bituminous water-proofing material and line the bottom of the planter area with 15-pound building felt or plastic sheeting to provide a moisture barrier. Then place 6 inches of sand or pea gravel for drainage, and cover that with a generous layer of topsoil.

Now put on your landscaper's hat and embellish your project with a variety of colorful plants.

Front-Entry Face-lift

If your home has a less-than-inspiring front entry, this project may be just the thing you've been looking for to bring your entry up to expectations. Don't let the prospect of building a masterful-looking arch frighten you—it's much easier than you'd expect.

When planning the project, make a scale drawing of the facade as you want it to look in your situation, taking care that you keep the arched coves in scale with their surroundings. Also pay close attention to the placement of the coves in relation to adjoining doors, walls, and other features of your home.

1 Start by reviewing the bricklaying basics discussed on pages 42–43, 50–51, and 60–65. Then, begin laying the bricks, and continue until you've completed the course on which the plywood arch template will rest.

Now build your template using ½-inch plywood and 2×3s as shown. Base the width of the form on the number of rowlock bricks—plus joints—you plan to use. The height and diameter is up to you.

After building the form, drylay the bricks around it and mark on the form the position of the mortar joints between the bricks that will curve around the form's top. Doing this now will make it easier to achieve even spacing when you actually lay the bricks.

2 Lift the form into position, then continue laying up each course of bricks until you reach the point where the arch starts to curve in. At that point, lay half-bricks over the curved form.

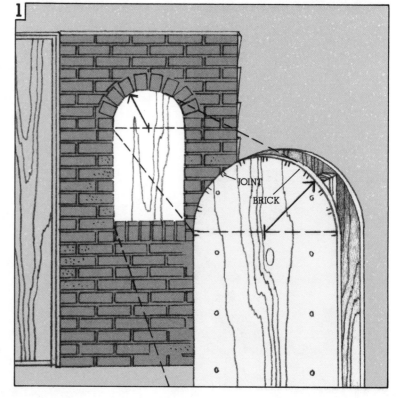

JOINT

BRICK

Cut adjoining units to fit (as shown) as you continue laying additional courses in the main section of the wall (see page 45 for details on cutting bricks). Use plenty of metal brick ties near the arch to anchor it to the existing wall studs. Nail the ties to the wall at stud locations, then bend them. They should span almost the entire width of the brick.

3 Allow the mortar in the arch and adjacent veneer to dry and cure for 24 hours before removing the plywood form. Nail expanded metal lath to the wall at the rear of the cove if you plan to finish it with stucco (see pages 68–69). Otherwise, cut a piece of exterior drywall to fit inside the arch, nail it to the wall, and apply two coats of the wall finish of your choice. If you stucco this area, apply materials carefully to avoid staining the bricks with mortar. After finishing the back surface, cut and lay the rowlocks to complete bricking of the cove.

4 Wrap up the project by preparing and installing appropriate trim to cover the seams where the veneer meets adjoining walls, soffits, and doors.

For a trim piece that fits the shape of lapped siding, use a spacer that matches the thickest part of the siding to scribe the material as shown. Cut it out carefully with a portable circular saw or saber saw and nail in place with galvanized hardboard siding nails.

Apply latex or butyl caulk to provide a weathertight seal where brick molding meets the veneer. Paint the caulk and new trim to complete the project. Later, when the mortar has cured, go back and scrub away any misplaced mortar or stains. (See page 73 for information on cleaning masonry with a mild acid solution.)

Corduroy-Block Privacy Screen

This charming courtyard is proof positive that with a little imagination you can transform an otherwise nondescript patio slab into something special. The corduroy-block walls surrounding the slab define the area beautifully and afford adequate privacy from outside distractions. Be aware, too, that several varieties of decorative block are available on the marketplace today. So if corduroy doesn't suit your taste, there are sure to be other styles that do.

If you don't yet have a patio to build around, see pages 10–11, 15–17, 20–21, 28–39, 42–47, and 50–53 for complete information on fashioning patios of concrete or brick, block, or stone.

When planning the placement of the walls, give yourself some room between the slab and the walls, as was done here, to allow for some landscaping flexibility. And before settling on a wall height—6 feet is typical—check with local building officials to see whether any codes restrict the height of such structures.

1 Begin by establishing the perimeters of the walls, excavating a trench twice the width of the wall thickness to below the frost line, and pouring concrete to within a few inches of grade. (See page 30 for help with determining the amount of concrete needed.) Allow the concrete to cure for several days, keeping it moist, if possible, for the duration.

Before proceeding, read pages 50–51 and pages 54–59 for information on mixing and working with mortar and on how to build concrete block walls to the line. Then lay the first course of blocks in a bed of mortar on the footing, staggering the joints in the two tiers as shown for better structural soundness.

Continue laying up the blocks, stacking those in each tier directly above the one in the previous course. To give the walls additional strength, insert metal ties every 2 or 3 feet along the walls and across the blocks, and in the corners of every third course. Inserting steel rerods down through the blocks' hollow cores every 6 to 8 feet of wall length and then filling the cores with mortar serves as still another structure-strengthening option.

2 When you reach finish-wall height, lay a course of solid concrete 4×8×16-inch blocks, as shown, to protect the wall from moisture problems, to give it a finished look, and to tie the tiers together at the top.

NOTE: If you wish, you can paint the blocks to complement the color of your house. If you go this route, check first with a paint dealer as to how best to prepare the blocks, as well as the type of paint to use.

4" OVERLAP

FOOTING TO BELOW FROST LINE

Rustic Stone Fireplace Facade

It used to be that if you wanted to enjoy the warmth and coziness of a fireplace in your home, you had to call in a mason to construct it. All that's changed now, however. The all-masonry fireplaces of days gone by are rarely built today. Why? Because they're too costly to construct, and generally they're not very energy-efficient, either. Today, what most people get is a prefabricated firebox and a metal chimney, both of which are concealed behind an attractive facing of masonry or some other material.

The fireplace featured here sports a lightweight stone veneer with brick border that any do-it-yourselfer can put up in a few hours, with professional-looking results. (See pages 66–67 for information on installing lightweight veneers.)

In any fireplace installation there's a lot of planning to do. First, you need to learn about the various units on the market to see which seems best for your situation. The more-efficient ones feature glass doors, ducting to supply combustion air from outdoors, and systems to distribute the warm air the fireplace generates.

Also, discuss your plans with the sales people at the retail outlet. They can be most helpful. And when the firebox and accessories arrive, study the installation instructions carefully. Pay particular attention to the minimum clearances between the firebox and chimney and combustible materials.

¾" PLYWOOD

2×8s @ 16" O.C.

2×8

24"

1 If the floor of the room you're heating is carpeted, pull the carpeting and pad back out of the way. Then construct a platform with 2x material and plywood as shown.

Now, with a helper or two, lift the fireplace up onto the platform. Install the ductwork and flue sections, then frame around the firebox with 2x stock. Cover the firebox surround with ½-inch drywall. Don't bother finishing the joints and corners of the sheetrock because the stone veneer will cover them.

2 Use roofing nails to fasten expanded metal lath to the vertical surfaces you plan to cover with stone veneer. Then after reviewing the material on pages 50–51, which tells how to mix and work with mortar, cut bricks in half and lay columns of bats along each side of the fireplace opening. Check often to ensure that the bricks in each column are plumb and in proper alignment. Embed metal brick ties in every fourth joint to anchor both columns to the wood framing behind them.

Place a 3×3×¼-inch steel angle of suitable length on top of the two columns. Now lay a course of bricks on the lintel as shown.

3 Before applying the lightweight stone, lay out the various pieces on the floor in a trial run. Doing this will help you fit them together in a suitable pattern and determine which pieces you must cut in order to fit.

When ready to start laying the veneer, mix and apply mortar to an area of about 4 square feet, beginning in an upper corner of the firebox. Press stones in the mortar as discussed on pages 66–67. Continue until all of the veneer is in place.

A day or so after the mortar has set, fill the joints between the stones with grout.

EXPANDED METAL LATH

ANGLE-IRON LINTEL

Wood-Stove Surround

Wood stoves radiate heat in all directions. That's why the installation instructions that accompany each unit specify that you must place it on a noncombustible surface and maintain a certain safe distance from other combustibles as well.

This stylish project gives effective, attractive heat protection and provides a handy storage cabinet where you can hide your woodburning accessories when not in use.

Before beginning, however, refer to the stove's installation instructions for help with determining the best location for your unit. And take note of the size of noncombustible pad and firewall the manufacturer recommends.

1 Begin by settling on the dimensions of the firewall, base, and cabinet. (Here, the firewall measures 54×54 inches; the base, 54×62 inches; and the cabinet 12×54×36 inches.) Then pull back the carpeting and pad from the area you'll be working in, and mark the outlines of the base and firewall.

Now remove a section of baseboard molding wide enough to accommodate both the firewall and the cabinet. With all this done, turn to page 43 for help with determining how much brick you'll need. (Remember to figure in your estimate the number of bricks needed to cover the back side of the cabinet.)

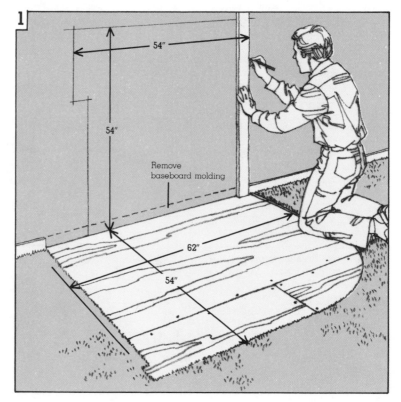

54"

54"

Remove baseboard molding

62"

54"

2 Begin construction of the cabinet by cutting the back, ends, and bottom from ¾-inch plywood. (You'll have to notch one end to fit around the baseboard molding.) Glue and nail these members together, making sure the bottom of the cabinet is 3¾ inches up off the floor. Then glue and nail 1×2 ledgers to the ends and back at whatever height you want to place the shelf.

With the cabinet lying on its back, glue and nail a 3¾-inch-wide length of 1x material on edge between the ends of the cabinet, 1¾ inches back from their front edge. Now, working on a flat surface, glue and nail together the 1×2 frame that will face the cabinet. Use flat metal T-braces to strengthen the joints.

Glue and nail the 1×2 frame to the cabinet. Then cut two plywood doors. Nail a 1×2 ledger to the center stile to support the shelf. Cut the shelf to size.

Now cut a ¾-inch plywood top the same length and width as the finished cabinet. Frame it with 1×3s, then cover both with plastic laminate.

Paint or stain and finish the cabinet, let it dry, then lower the shelf into place and the laminate-covered top down onto the cabinet. Hang the cabinet doors with self-closing hinges. Set the cabinet in place, and secure it to the wall and floor.

3 Lay the firewall's border bricks first, then the bricks within that border, then those on the cabinet back (see pages 66–67 for installation particulars). Now arrange the border for the floor section in a dry run to establish joint thickness before spreading adhesive and laying them permanently. Then, starting in a corner, lay the remainder of the bricks for the base. Cut bricks to fit as needed. Once all bricks are in place, grout the joints. Cut the carpeting and pad to fit, then tack it in place.

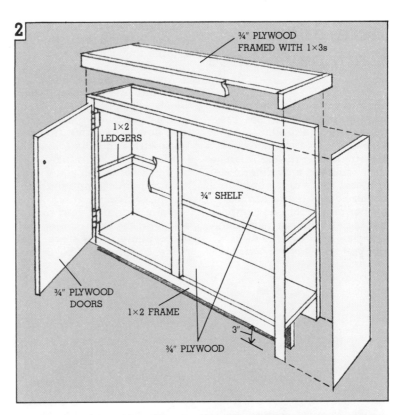

2

¾" PLYWOOD FRAMED WITH 1×3s

1×2 LEDGERS

¾" SHELF

¾" PLYWOOD DOORS

1×2 FRAME

¾" PLYWOOD

3"

3

Real-Brick Chimney Chase

With today's all-metal chimneys, safely venting hot flue gases from a fireplace or a woodburning stove doesn't require nearly the skill it once did. You need only determine the best route to the outside of the house, then snap together manufactured components until the flue is a specified height above the roof. But you still have to conceal your handiwork somehow. And unless you were able to run the flue straight up through the roof, you'll have to build a chimney chase.

On this and the next page, we show you not only how to frame a chimney chase, but also how to face that framework with handsome, longlasting brick. We think you'll be surprised when you discover how "accomplishable" this project is.

When planning your project, read the instructions that accompanied your stove, fireplace, or furnace. They specify minimum clearances between the flue and combustible materials, as well as height guidelines to ensure a positive draft situation.

Also review pages 30 and 42–43 about estimating your concrete and brick needs, and pages 50–51 and 60–65 about working with mortar and erecting brick walls.

And when determining the dimensions of the chase, keep in mind that the wider the chase (within reason) the more impressive the structure. Five-foot-wide chases are not at all uncommon.

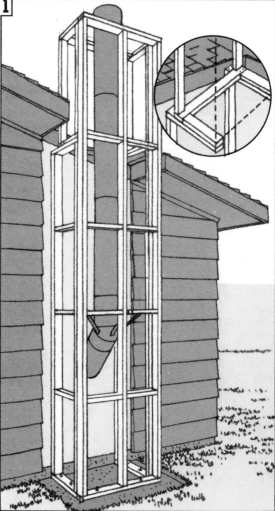

1 Start by excavating far enough down to provide a frost-free pad footing to support the framing and brick veneer. Once the excavation is complete, insert lengths of reinforcing rod into the existing block or poured concrete wall (see page 29 for details) to tie the footing to the foundation. Then fill the excavation to within a few inches of grade with concrete. Set anchor bolts in the concrete to help tie the chase to the concrete. Allow the footing to set up for several days.

Meanwhile, cut a large enough opening in the roof overhang for the chimney chase and brick veneer to pass through it. Snap vertical chalk lines where the brick veneer will meet the wall of the building. Saw along these lines just deep enough to cut the siding and remove it.

Build the wooden chase as shown, nailing to the building's framing members wherever possible. Note in the detail how at the roof line the back wall of the chase is recessed. You must do this to provide support for bricks above this point. Install the metal flue and cover the frame with insulated sheathing.

2 Lay brick veneer up the faces of the chase, tying the two together with metal ties every 16 inches. (If necessary, refer again to pages 50–51 and 60–65 for advice on laying bricks.) When you reach the roof line (you'll need scaffolding to complete the project), position the steel angle lintel as shown in the detail. And remember to put in metal flashing as the courses pass the roof line to help safely channel moisture away from the chimney.

3 In the fifth course from the top of the chimney, begin creating a *corbel* effect by laying the bricks so they overlap the previous course by ½ inch. Lay the next three courses in the same manner, cutting brick pieces to fill the voids that result, as shown in the detail.

And bring the final course back in line with earlier courses to tie the corbeled courses together.

4 Top off the galvanized metal chimney with a custom-made chase top. (Order this from a local sheet-metal shop. When placing your order, have ready the outside dimensions of the chase top, plus the size and location of the hole for the flue. Or, if you prefer, hire a qualified person to do this for you.) After placing the top, install the storm collar and flue cap as shown. Nail the chase top to the brick with masonry nails and caulk around the edges.

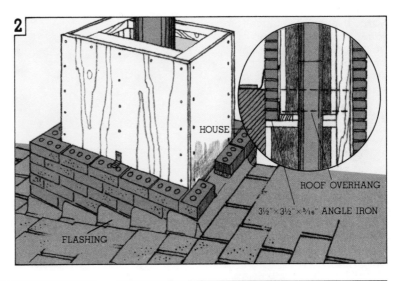

HOUSE

ROOF OVERHANG

3½" × 3½" × ⁵⁄₁₆" ANGLE IRON

FLASHING

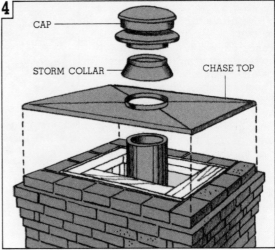

CAP

STORM COLLAR

CHASE TOP

Center-stage Barbecue Center

For many chefs, preparing great-tasting food outdoors is a delicate balance of know-how and showmanship. And what could be a more fitting stage to perform culinary feats on than this sleek, contemporary barbecue center.

Simply styled and not difficult to construct, this project will make a handsome, long-lasting addition to any backyard. If possible, place it adjacent to or in a corner of your patio so you can enjoy chatting with family or friends while keeping an eye on your sizzling steaks.

1 Stake out the perimeters of the project and excavate for footings that reach below the frost line. Make the perimeter for the barbecue pit's footing approximately 3×3 feet and the trenches for the wall footings about 8 inches wide. Providing the soil is firm enough, you can use the walls of the excavation as the form for the concrete.

From your concrete supply dealer, buy a 22-inch-diameter cardboard form tube long enough to reach from a point 26½ inches above grade to the bottom of the excavation. Set the tube into the center of the 3×3-foot excavation. To keep it from moving, stabilize it at ground level as shown here. Then after checking to make sure the tube is plumb, fill it with sand to a level about 2 inches below final grade. Carefully pour concrete around the base of the tube and fill the trenches to complete construction of the footings.

2 Allow the concrete to cure for several days. Review the material about estimating and bricklaying on pages 43, 50–51, and 60–65, then begin laying up half-bricks,

cut ends out, around the cardboard tube. (See pages 46–47 for how to cut bricks.) Be sure to leave an opening for the ash door. Note the ⅜″×2″ horizontal steel bar used to support the bricks above the opening.

After completing the seventh course, drill 3 equally spaced ⅜-inch holes through the tube to accommodate the ⅜-inch metal grate. Insert the bent rods as shown and lay two more courses. Insert 3 more rods to hold the grill and lay the final course of bricks.

3 Using a running bond, lay the first four courses of the outside of the barbecue pit and the low walls adjoining it. Position the cast-iron ash door at the appropriate place as you do so. For help with how to tie the two-tier low walls to the walls of the pit, see page 65.

Now build and position temporary 2×4 wood supports as was done here. Cantilever the fifth course out onto the wood supports, then refer to the detail for positioning the remaining courses. Remove the temporary wood supports after the mortar has set and cured for 24 hours.

4 Peel the cardboard form from the inside of the pit all the way down to the sand. Smooth the sand and cover it with a 4-inch layer of concrete (see the detail). Slope the surface slightly toward the opening so the pit will drain.

Fill the voids between the circular wall and the outside wall of the barbecue with concrete. Install the grate and grill. Allow the whole project to cure for 10 days or so before using it. Premature use may crack the mortar joints.

⅜″ GRILL SUPPORT

⅜″ GRATE SUPPORT

⅜″×2″ STEEL BAR

FIRE PIT

TEMPORARY WOOD SUPPORTS

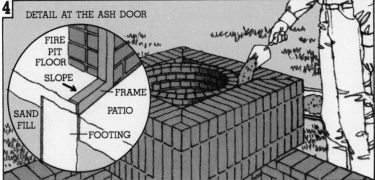

DETAIL AT THE ASH DOOR

FIRE PIT FLOOR

SLOPE

FRAME

PATIO

SAND FILL

FOOTING

Down-Home Stone Fence

This fencing idea offers the best of both worlds—it visually establishes a boundary, yet leaves unrestricted the view for those on either side of it. And because the rails are not joined permanently to the stone posts with mortar, replacing them—if necessary—is a snap.

When planning this project, check with local authorities to see what building codes apply to fences and their location in relation to your property line. Also carefully plot the location of the posts. If you plan to use 8-foot-long rails, for instance, you'll want the facing walls to be about 7½ feet apart to provide support for each end of the rails.

1 Stake the corners of the fence. Then using a mason's line that is stretched between the corner stakes, stake the four corners of each post. To ensure long-term stability of the posts, excavate to below the frost line. (If you plan to incorporate an outdoor light as was done here, dig a 12-inch-deep trench for the electrical wiring, position the light pole, run wiring to it, and stabilize the pole at ground level with bracing.) Pour concrete into each of the excavations to within an inch of grade. See pages 30–31 for estimating and ordering or mixing concrete.

2 After reviewing the material on pages 33 and 48–51, lay up the ashlar stone posts. Make the posts anywhere from 42 to 48 inches tall. Use large pieces of stone to cap off each post. Leave the insides of all posts hollow (except those supporting the light fixture) and insert the rails in the openings.

CONDUIT 12" BELOW GRADE

No mortar at rail to allow for expansion and replacement

Glossary

The following definitions help explain the terminology of masons and concrete workers. For further explanations, and for words not listed here, refer to index (pages 95-96).

Aggregate—Gravel or crushed rock that when mixed with sand, portland cement, and water forms concrete.

Ashlar—Rectangular blocks of stone of uniform thickness used mainly to build dry walls. (See also *dry wall*.)

Backfill—The soil used to fill in an excavation next to a wall. This soil adds stability to the wall and keeps water away from it.

Backplastering—Covering a masonry surface with (usually) two thin layers of mortar for the purpose of sealing it against moisture. Typically, a layer of bituminous waterproofing material is applied after the mortar sets up. Also known as *parging*.

Bat—A half-brick. Bats are used when whole bricks won't fit into the alloted space.

Batter—The practice of tapering the sides of a stone wall to give it added stability.

Batter boards—A board frame supported by stakes set back from the corners of a structure that allows for relocating certain points after excavation. Saw kerfs in the boards indicate the location of the edges of the footings and the structure being built.

Bed joint—The layer of mortar between two courses of masonry units. (See also *course*.)

Bond—(1) Any one of several patterns in which masonry units can be arranged. (2) To join two or more masonry units with mortar.

Brick set—A wide-bladed chisel used for cutting bricks and concrete blocks.

Buttering—Smearing mortar on bricks or blocks with a trowel before laying them.

Chink—A narrow piece or sliver of stone driven into cracks or voids in a stone wall to achieve added stability.

Closure brick (or block)—The final unit laid in a course of bricks (or blocks).

Concrete nails—Hardened steel nails that you can drive into solid concrete.

Control joint—A groove tooled into a concrete slab during finishing to prevent uncontrolled cracking later on. These joints, to be effective, should be one-fourth the thickness of the slab.

Coping—A brick, block, stone, or concrete cap placed at the top of a masonry wall to prevent moisture from falling directly on it and weakening the wall.

Corbel—A ledge or shelf formed by laying one or more courses of masonry units so they protrude from the face of a wall.

Corner lead—The first few courses of masonry laid in stairstep fashion at a corner to establish levels for the remaining units in those courses.

Course—A row of masonry units. Most projects consist of several courses laid atop each other and separated by mortar.

Cupping the mortar—Rolling a slice of mortar for the purpose of loading it onto a mason's trowel.

Darby—A long-bladed wood float commonly used to smooth the surface of freshly poured concrete in situations where using a smaller float isn't practical.

Dry wall—A wall of masonry units laid without mortar.

Edger—A concrete finishing tool for rounding and smoothing edges, which strengthens them.

Efflorescence—A powdery stain, usually white, on the surface of or between masonry units. It is caused by the leaching of soluble salts to the surface.

Expansion joint—The vertical space built into a structure or between it and an existing structure to allow the concrete to expand and contract with temperature changes without damage to the surface.

Exposed aggregate—A concrete finish achieved by embedding aggregate into the surface, allowing the concrete to set up somewhat, and then hosing down and brushing away the concrete covering the top portion of the aggregate.

Face brick—A type of brick made specifically for covering (veneering) walls.

Finish coat—The final coat of mortar or plaster in a stucco finish. (See also *stucco*.)

Flashing—A layer of material, usually metal, inserted in masonry joints and attached to adjoining surfaces to seal out moisture. Typically, a layer of bituminous material applied at the joint of the two surfaces completes the seal.

Glossary *(continued)*

Float—A rectangular wood or metal hand tool that is used for smoothing and compressing wet concrete.

Footing—A thick concrete support for walls and other heavy structures. To ensure against damage from frost heave, footings should extend to below the frost line and be on firm soil.

Frost line—The maximum depth frost normally penetrates the soil during the winter. This depth varies from area to area depending on the climate.

Grout—A thin mortar mixture. (See also *mortar*.)

Hawk—A fairly small board with a handle beneath it used for holding mortar.

Head joint—The layer of mortar used to tie the ends of adjoining masonry units together.

Jointer—A tool used for making grooves or control joints in concrete surfaces to control cracking. (See also *control joint*.)

Joint strike—A tool used to finish the joints between masonry units. Joints are struck for aesthetic reasons as well as to compress the mortar into the joints.

Masonry cement—A special mix of portland cement and hydrated lime used for preparing mortar. The lime adds to the workability of the mortar.

Modular spacing rule—A measuring device used to verify that a course of masonry units is at the proper height.

Mortar—A mixture of masonry cement, masonry sand, and water. For most jobs, the proportion of cement to sand is 1:3.

Nominal dimensions—The actual dimensions of a masonry unit, plus the thickness of the mortar joints on one end and at the top or bottom.

Parging—See *backplastering*.

Plumb—The condition that exists when a surface is at true vertical.

Plumb bob—Weight used with a plumb line to align vertical points.

Pointings—See *Tuck-pointing*.

Pre-mix—Any of several packaged mixtures of ingredients used for preparing concrete or mortar.

Ready-mix—Concrete that is mixed as it is being trucked to the job site.

Re-rod (reinforcing rod)—Steel rod that is used for reinforcing concrete and masonry structures.

Retaining wall—A wall constructed to hold soil in place.

Rowlock course—Several bricks laid side by side on their faces and pitched slightly (when used outside) to shed moisture. Typically, rowlocks are used below windows and as caps on walls.

Rubble—Uncut stone found in fields or as it comes from a quarry. Often used for dry walls.

Scratch coat—The first coat of mortar or plaster, roughened so the next coat will stick to it.

Screed—A straightedge, often a 2×4 or 2×6, used for leveling concrete as it is poured into a form.

Set—The process during which mortar or concrete hardens.

Slump—The wetness of a concrete or mortar mix; the wetter the mix, the more it spreads out or slumps.

Story pole—A measuring device, often a straight 2×4, with a series of marks set at regular intervals, used to verify that a course of masonry units is at the proper height.

Stretcher—A brick or block laid between corner units.

Striking iron—See *joint strike*.

Stucco—A finish comprised of two or more layers of mortar (white or colored) applied to either indoor or outdoor walls.

Throw mortar—To place mortar using a mason's trowel.

Transit-mixed concrete—See *ready-mix*.

Trowel—Any of several flat and oblong or flat and pointed metal tools used for finishing and/or handling concrete and mortar.

Tuck-pointing—The process of refilling old joints with new mortar.

Veneer—A layer or bricks or stones that serves as a facing.

Weep holes—Openings made in mortar joints to facilitate drainage of built-up moisture.

Welded wire fabric—A steel screening used to reinforce certain types of concrete projects such as walks, drives, and patios.

Index